SpringerBriefs in Applied Sciences and Technology

SpringerBriefs present concise summaries of cutting-edge research and practical applications across a wide spectrum of fields. Featuring compact volumes of 50 to 125 pages, the series covers a range of content from professional to academic.

Typical publications can be:

- A timely report of state-of-the art methods
- An introduction to or a manual for the application of mathematical or computer techniques
- A bridge between new research results, as published in journal articles
- A snapshot of a hot or emerging topic
- An in-depth case study
- A presentation of core concepts that students must understand in order to make independent contributions

SpringerBriefs are characterized by fast, global electronic dissemination, standard publishing contracts, standardized manuscript preparation and formatting guidelines, and expedited production schedules.

On the one hand, **SpringerBriefs in Applied Sciences and Technology** are devoted to the publication of fundamentals and applications within the different classical engineering disciplines as well as in interdisciplinary fields that recently emerged between these areas. On the other hand, as the boundary separating fundamental research and applied technology is more and more dissolving, this series is particularly open to trans-disciplinary topics between fundamental science and engineering.

Indexed by EI-Compendex, SCOPUS and Springerlink.

More information about this series at http://www.springer.com/series/8884

Adnan Bakri · Mohd Al-Fatihhi Mohd Szali Januddi

Systematic Industrial Maintenance to Boost the Quality Management Programs

Springer

Adnan Bakri
Advance Facilities Maintenance
Engineering Technology (AFET)
Research Cluster, Facilities Maintenance
Engineering Section, Malaysian Institute
of Industrial Technology
Universiti Kuala Lumpur
Johor Bahru, Johor, Malaysia

Mohd Al-Fatihhi Mohd Szali Januddi
Advance Facilities Maintenance
Engineering Technology (AFET)
Research Cluster, Facilities Maintenance
Engineering Section, Malaysian Institute
of Industrial Technology
Universiti Kuala Lumpur
Johor Bahru, Johor, Malaysia

ISSN 2191-530X   ISSN 2191-5318   (electronic)
SpringerBriefs in Applied Sciences and Technology
ISBN 978-3-030-46585-8   ISBN 978-3-030-46586-5   (eBook)
https://doi.org/10.1007/978-3-030-46586-5

This Springer imprint is published by the registered company Springer Nature Switzerland AG
The registered company address is: Gewerbestrasse 11, 6330 Cham, Switzerland

Acknowledgements

This book would not be possible without the support of Facilities Maintenance Engineering (FAME) section members and Advance Facilities Engineering Technology (AFET) research cluster from Universiti Kuala Lumpur Malaysian Institute of Industrial Technology (UniKL MITEC).

Contents

Chapter 1
Introduction

1.1 Overview of Quality Management Programs

In today's highly competitive business environment, quality emerges as an effective strategy for manufacturing companies toward the success, growth, and enhances their competitive position. In order to survive, every manufacturing company has to infuse quality improvement (QI) initiatives such as total quality management (TQM), lean manufacturing (Lean) and just-in-time (JIT) in all aspects of their operations. The emergence of progressive QI initiatives has changed tremendously the nature of manufacturing environment. They are utilized with the aim at positioning ahead of competition in term of production efficiency, excellent product quality, meet customers' deadline, and optimize the operational cost.

1.2 Total Quality Management

Under TQM philosophy, the quality control and assurance of product is moved to the production process instead of inspection at the final product. Defects and variation of product are eliminated at the production processes through adequate process control techniques. A core definition of TQM outlines a management approach to long-term success through customer satisfaction. TQM requires a full participation from all people in the hierarchy of an organization toward improving the quality of both products and services in which they work. The operational approach of TQM is based on the eight principles:

- **Customer-focused:** The main driving force of TQM is customer satisfaction. Customer-focused is the orientation of an organization toward fulfilling the customer's requirement. The importance of customer focus cannot be under-estimated; it is the major determinant of the success or failure of an organization.

© The Author(s), under exclusive license to Springer Nature Switzerland AG 2020
A. Bakri and Mohd. A.-F. Mohd Szali Januddi, *Systematic Industrial Maintenance to Boost the Quality Management Programs*, SpringerBriefs in Applied Sciences and Technology,
https://doi.org/10.1007/978-3-030-46586-5_1

- **Total employee involvement:** All employees in the organization hierarchy requires to participate and work hand in hand to satisfy the customer requirement.
- **Process-centered:** TQM focuses on the continual improvement on processes or system used in the organization. A process is a series of steps to produce either products or services in the organization. In the case of manufacturing organization, the processes and systems start from the inputs from suppliers (internal or external) and transform them into outputs that are delivered to customers (internal or external). The processes and system are required to be continuously monitored in order to detect unexpected variation.
- **Integrated system:** TQM refers to an integrated approach by management to focus all functions and levels of an organization's hierarchy on quality and continual improvement.
- **Strategic and systematic approach:** A critical part of the management of quality is the strategic and systematic approach to achieving an organization's vision, mission, and goals. This process, called strategic planning or strategic management, includes the formulation of a strategic plan that integrates quality as a core component.
- **Continual improvement:** A large aspect of TQM is continual process improvement. Continual improvement drives an organization to be both analytical and creative in finding ways to become more competitive and more effective at meeting the customer requirements.
- **Fact-based decision-making:** In order to know how well an organization is performing, data on performance measures is necessary. TQM requires that an organization continually collects and analyzes data in order to improve decision-making accuracy, achieve consensus, and allow prediction based on past history.
- **Communications:** Effective communications plays a large part in maintaining morale and in motivating employees at all levels in the organization's hierarchy. Communications involve strategies, method, and timeliness.

1.3   Just-in-Time (JIT)

JIT is one the key elements in lean manufacturing. It emerges as a technique for inventory management. The principle that underpins JIT is that production should be "pulled through" rather than "pushed through." This means that product is produced based on specific orders from customer. The production starts once a customer has placed an order with the manufacturer. JIT, however, is not a mere inventory control technique, but a manufacturing system that tries to enhance quality and lower costs through the reduction of inventories and shortening lead times. JIT system is designed to operate in an ideal environment such as with constant processing times, smooth and stable demand, and uninterrupted processing. However, in a real-life environment, the success of JIT system is subjected to various uncertain factors including unstable processing times, variable demand, and process interruption due to maintenance issues.

1.4 Lean Manufacturing

Lean manufacturing is a philosophy and a way of working aims at reducing the operating costs through the elimination of waste in the production processes. Lean principles have greatly influenced manufacturing as well as service industry worldwide. The benefits of lean include reduced lead times, reduced operating costs, and improved product quality, to name just a few. "Waste" is defined as anything that does not add value in the production process. Lean manufacturing focuses on the elimination of waste in all forms, these include: unnecessary transportation;

- Excess inventory;
- Unnecessary motion of people, equipment, or machinery;
- Waiting, whether it is people waiting or idle equipment;
- Over-production of a product;
- Over-processing or putting more time into a product than a customer needs, such as designs that require high-tech machinery for unnecessary features; and
- Defects, which require effort and cost for corrections.

1.5 Five Principles of Lean Manufacturing

A widely referenced book, Lean Thinking: Banish Waste and Create Wealth in Your Corporation, laid out five principles of lean, which many in the field reference as core principles. They are value, the value stream, flow, pull, and perfection. These are now used as the basis for lean implementation.

i. **Identify value from the customer's perspective.** Value is created by the producer, but it is defined by the customer. In other words, companies need to understand the value the customer places on their products and services, which, in turn, can help them determine how much money the customer is willing to pay. The company must strive to eliminate waste and cost from its business processes so that the customer's optimal price can be achieved at the highest profit to the company.

ii. **Map the value stream.** This principle involves recording and analyzing the flow of information or materials required to produce a specific product or service with the intent of identifying waste and methods of improvement. The value stream encompasses the product's entire lifecycle, from raw materials through to disposal. Companies must examine each stage of the cycle for waste—or muda in Japanese. Anything that does not add value must be eliminated. Lean thinking recommends supply chain alignment as part of this effort.

iii. **Create flow**. Eliminate functional barriers and identify ways to improve lead time to ensure the processes are smooth from the time an order is received through to delivery. Flow is critical to the elimination of waste. Lean manufacturing relies on preventing interruptions in the production process and enabling a harmonized and integrated set of processes in which activities move in a constant stream.

iv. **Establish a pull system.** The new work should be started when there is demand for it. Lean manufacturing uses a pull system instead of a push system. With a push system, inventory needs are determined in advance and the product is manufactured to meet that forecast. However, forecasts are typically inaccurate, which can result in swings between too much inventory and not enough, as well as subsequent disrupted schedules and poor customer service. Pull relies on flexibility and communication.

v. **Pursue perfection with continual process improvement, or kaizen**. Lean manufacturing rests on the concept of continually striving for perfection, which entails targeting the root causes of quality issues and ferreting out and eliminating waste across the value stream.

1.6   Seven Lean Manufacturing Tools and Concepts

Lean manufacturing requires a relentless pursuit of reducing waste. Waste is anything that customers do not believe adds value and for which they are not willing to pay. This requires continuous improvement, which lies at the heart of lean manufacturing. Other important concepts and processes lean relies on include:

i. **Heijunka:** production leveling or smoothing that seeks to produce a continuous flow of production, releasing work to the plant at the required rate and avoiding interruptions.

ii. **Kanban:** A signal—either physical, such as tag or empty bin, or electronically sent through a system—used to streamline processes and create just-in-time delivery.

iii. **Jidoka**: A method of providing machines and humans with the ability to detect an abnormality and stop work until it can be corrected.

iv. **Andon:** A visual aid, such as a flashing light, that alerts workers to a problem.

v. Poka-yoke: A mechanism that safeguards against human error, such as an indicator light that turns on if a necessary step was missed, a sign given when a bolt was tightened the correct number of times or a system that blocks a next step until all the previous steps are completed.

vi. 5S: A set of practices for organizing workspaces to create efficient, effective, and safe areas for workers and which prevent wasted effort and time. 5S emphasizes organization and cleanliness.

1.7 Overview of Industrial Maintenance

There is consensus among many authors that QI initiatives are influenced by equipment reliability and maintainability. The wastes generated in production have a strong relationship with the performance of production equipment. The malfunction and breakdown of equipment would result in poor quality product and as a consequence delayed delivery. Satisfying customers' requirement in timely manner means the equipment availability is required to be at its peak level. Through a systematic and strategic maintenance management, defects and variations resulting from the poor equipment could be eliminated. Hence, equipment effectiveness is no longer restricted to availability, but involves other factors, such as quality and efficiency. However, it is observed that the significant role of maintenance management as an important constituent of TQM, lean manufacturing, and JIT is not emphasized in the literature. Ineffective maintenance management will have significant impact on company's profitability. A huge amount on maintenance of inefficient production equipment would increase the company operational costs. Effective maintenance management of production equipment and system is one of the vital requirements toward achieving world-class manufacturing. Maintenance management is considered as "the last frontier" for manufacturing facilities. In this sense, reliable equipment is considered as the main elements toward performance as well as profitability of the organization.

It is imperative that those QI initiatives to be integrated with maintenance management. Other literatures also provide evidences of some ineffectiveness in maintenance management in manufacturing industries particularly in small firms, such as too much "fire-fighting," i.e., still applying the concept of breakdown maintenance and limited preventive approaches. Most of these issues were nevertheless pointed out by research on manufacturing firms in several western countries and local industries. Despite those notes, there were also some strong points brought to light, including the growing awareness on the importance of maintenance management and the increasing diffusion of latest maintenance management and technology across all industries, such as CMMS, RCM, CBM, and TPM. Without any doubt, maintenance management plays an important role to spur the company's performance, particularly as part of essential strategies to face the competition and cope up with increasing pressure of market globalization. As such, maintenance management required a continual study to further improve its effectiveness.

1.8 Trends in Industrial Maintenance

The concepts of maintenance management have evolved over the last few decades. The development of maintenance management research can be classified into two major trends as illustrated in Fig. 1.1, namely:

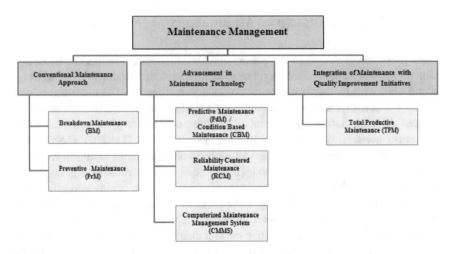

Fig. 1.1 Concept in maintenance management

- Conventional maintenance: Breakdown-based and preventive maintenance.
- Advancement in maintenance technology and maintenance methods. For instance: the emergence of CMMS, RCM, and CBM.
- Integration of maintenance activity with QI initiatives. For instance: development of TPM as a vital subset of TQM.

Chapter 2
Conventional Maintenance Approach

2.1 Breakdown Maintenance

Breakdown maintenance (BM) focuses on repair and restoration action performed on equipment once the failure occurred. The aim of BM is to bring failed equipment to at least at its minimum acceptable condition. It is the oldest practice of maintenance management and is executed at unpredictable period since the equipment or system's failure time is not known. This concept has the drawback of unplanned breakdown, in which might resulted in excessive damage of parts, inventory problem on spare parts, increase in repairing and troubleshooting time and cost. BM is also referred to as maintenance or engineering work related to unforeseen plant breakdowns. It is not a planned event and as such can cost the factory lost production and sales as well as other expenses such as out of budget maintenance costs including overtime, technician call outs and urgent delivery fees for spare parts or support. Maintenance breakdowns can have many causes, some being the result of improper preventive and long term maintenance planning, lack of plant inspections, lack or incorrect evaluation of stresses and load cycles on machinery, faulty design and materials and in some cases neglect.

A proper maintenance plan, correct documentation, record keeping and maintenance execution at the correct intervals usually aids the maintenance or engineering department to eliminate the majority of unforeseen breakdowns. Unforeseen plant breakdowns can be classified as a form of waste in the lean manufacturing system as they waste both time and resources. This indirectly costs the business financially as budgeted maintenance costs are used for unforseen costs and work as well as lost production which is lost revenue for the business. It is therefore important to eliminate or reduce BM to minimal levels to maximize production output and operational costs.

© The Author(s), under exclusive license to Springer Nature Switzerland AG 2020
A. Bakri and Mohd. A.-F. Mohd Szali Januddi, *Systematic Industrial Maintenance to Boost the Quality Management Programs*, SpringerBriefs in Applied Sciences and Technology, https://doi.org/10.1007/978-3-030-46586-5_2

2.2 Preventive Maintenance Techniques

Preventive maintenance (PM) primary objective is to prevent the failure of equipment before it occurs. In other word, PM aimed at the prevention of spontaneous breakdowns and failures. It is performed based on scheduled and planned basis. This concept was introduced in 1950s. It relies on the prediction that the equipment will be subjected to deterioration in performance once it reached a specified usage time. World class maintenance programs embrace a proactive approach, in which PM represents nearly 60% or more compared to other maintenance activities. Effective PM programs reflect an organization's best practices of planning, coordination, evaluation and continuous process improvement throughout an organization. With a strong emphasis and systematic PM program, plant effectiveness, reliable operations and lifetime usage of the equipment are greatly increased.

2.3 Historical Perspective of PM

The first industrial revolution began in the early 1800s. At that time, the industrial sector started to utilize the water and steam-powered engines for productivity and efficiency purposes. Prior to that, the production processes were performing manually, depending on either human or animal power to generate mechanism and power. In the late 1800s, the world entered a second industrial revolution (also known as electrical revolution) with the introduction of electricity in the manufacturing industry. Electrical energy was then used as a primary source of power. Electrical machines were more efficient to operate and maintain compared to water and steam based machines. The concept of mass production with multiple assembly and conveyor line were introduced as a way to boost productivity, efficiency, cost and resources optimization.

In the early 1970s, a third industrial revolution were slowly taken place as the automation and computerization processes have continuously evolved with the advances in the electronics and IT industry. Machines and equipment design in the third industrial revolution were generally quite complicated and relatively fast running with advance instrumentation and control systems compared the one in the previous era. In this era, the manufacturers were having a stiff competition in the market place particularly to: fulfill the on time delivery to customer; satisfy customer by producing a defect free products; maintain and optimize the overall operational costs. Due to such stringent requirements, the equipment and manufacturing machines are required to be at superb availability, so that downtime became a critical issue and it was inadequate to maintain on a breakdown basis. Manufacturing industry demanded a better maintenance and no intolerance of downtime which lead to the development of Planned or Preventative Maintenance.

2.4 PM Philosophy

PM is applied to most mechanical equipment. It is the most fundamental maintenance techniques. PM encompasses periodic inspection and the implementation of remedial steps to avoid unanticipated breakdowns and stoppages on production equipment. The typical PM work may include inspection and minor maintenance activities on equipment such as parts cleaning (C), lubrication (L), tightening (T), and minor refurbishment, replacement or adjustment (M). The key for effective PM is on organization's commitment to strictly follow the plan, schedule for CLTM based on a routine basis.

2.5 Operational Approach of PM

PM tasks can also be defined as a list of tasks for maintenance activities (CLTM) based on Original Equipment Manufacturer (OEM) recommendations. The OEM commonly provides complete recommendations for the installation, operation, inspection, and preventive maintenance of its machine or equipment. These recommendations should be used as the basis of a fundamental PM program.

2.6 Example of Implementation Step for PM

Without a doubt, developing and implementing a PM program takes time and energy. However, once it is established with staff trained to execute it, the benefits of PM far prevail over the costs associated with reactive or emergency maintenance that often results in unforeseen downtime, equipment replacement, and operation disruption. Having a PM in place that monitors company assets makes it possible for flexible maintenance scheduling saving time, money and energy. Below are the possible steps to develop the PM plan and schedule.

2.6.1 Step 1: Plan: Select the Equipment

PM procedures can be determined based on: (1) Referencing the OEM's manuals - for new equipment, or: (2) prior corrective maintenance recorded and experiences – for used or older equipment. For this book, the authors will focus on a new equipment only. OEM have a plethora of statistical data from in-house testing and field tests done by customers. The manuals they provide often contain schedules for necessary maintenance, the usage of critical spare parts, and basic maintenance work instructions. By looking at the OEM recommendations, write down the list of tasks and its

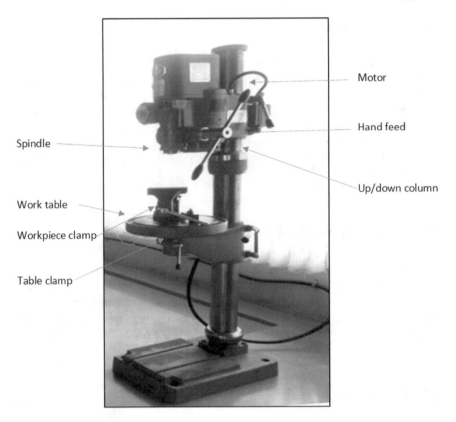

Fig. 2.1 Drilling machine and its part list

respective frequency (i.e., weekly, monthly, quarterly, semi-annually, annually). It is important to make note of these different scheduling scenarios while also estimating how much time may be needed to perform the PM. An important part of creating PM checklist is making a list of tools and internal and external resources needed to complete each job. In summary, a PM plan should include the following: a parts list, standard operating procedures (SOPs), required tools and estimated time to complete the PM tasks (Fig. 2.1; Table 2.1).

2.6.2 Step 3: Do: Train the Maintenance Staff

It is essential that the organization proceed the training of its maintenance staff as they are the core implementer of PM program. Having staff members trained to execute PM program is a key determinant of successful outcomes.

Table 2.1   A typical PM checklist for drilling machine

Preventive Maintenance Checklist

Dept	: A&T	Area	: Warehouse A2
Section	: FAME	Machine	: Drilling Machine -AP2018
Date	: 22/12/2019	Approval	
HOD	: Abu Zar		

No	Machine parts	Task	Daily	Weekly	Monthly	Every 6 months
1	Spindle	Greasing / check rotation accuracy				↗
2	Work table	Cleaning and remove chips/dirts	↗			
3	Workpiece clamp	Greasing	↗			
4	Table clamp	Tightening	↗			
5	Motor	Inspect for damage to wiring			↗	
6	Hand feed	Cleaning / tightening			↗	
7	Up/down column	Cleaning / greasing			↗	
8	Switch (ON / OFF)	Check operation of start/ switch			↗	
9	Emergency stop button	Check functioning of emergency stop		↗		

Tools :

Mechanical tools : Common spannar set

Electrical tools   : Ammeter

General tools   : Scrapper, cotton rags, grease.

2.6.3   Step 4: Check: Analyze–Adjust–Improve

It is paramount important to analyze the results of a PM program and adjust or improve it as needed. PM programs help organization identify equipment that require more time and money than others, leading to adjustments in the PM procedure/schedule.

References

1. General Information (1) on Preventive Maintenance (2019) Available at: https://www.hippoc mms.com/blog/preventive-maintenance-program-in-six-steps. Retrieved date: 13 March 2019
2. General Information (2) on Preventive Maintenance (2019) Available at: https://www.plantserv ices.com/articles/2018/maintenance-management-dont-let-small-problems-combine-into-a-fai lure-event. Retrieved date: 20 March 2019
3. Bloch HP, Fred Geitner: "Machinery Component Maintenance and Repair" (2004) 3rd Ed., Elsevier Publishing Company, NY and London, ISBN 0-87201-781-8

4. Bakri A, Rahim AR, Yusof M (2014) Maintenance management: Rationale of TPM as the research focus. In: Applied Mechanics and Materials Vol. 670. Trans Tech Publications Ltd, pp 1575–1582

Chapter 3
Advancement in Maintenance Technology

3.1 Predictive Maintenance

The aim of predictive maintenance (PdM) or often referred as condition-based maintenance (CBM) is to perform a scheduled maintenance before failure occurred on the equipment. CBM is carried out upon availability of validated and analyzed historical data on the performance and physical condition of equipment such as vibration, humidity, temperature, particle content in lubricant. The diagnostic techniques are used to prepare an appropriate maintenance plan. The typical CBM technologies used to diagnose equipment condition includes infrared thermography, vibration analysis, acoustic level measurements, oil particle analysis, electrical current test, and other designated tests. CBM dictates that maintenance should only be performed when certain indicators show signs of decreasing performance or upcoming failure. CBM is a superior method of equipment maintenance based on using real-time data to prioritize and optimize maintenance resources. CBM is a subset of RCM.

CBM has a number of benefits. For starters, because it works while the equipment is in service, it does not interrupt equipment operation. It can help ensure equipment reliability and worker safety and also reduce failure rates and unscheduled downtimes. Moreover, because maintenance activities are scheduled ahead of time, CBM tends to be less costly than preventive maintenance. This method has some limitations as well though. The tools used to monitor equipment for CBM can be expensive to install. Employees must be trained to use CBM technology effectively, which can cost time and money. Furthermore, the sensors employed might not work in harsher operating environments and can have trouble detecting fatigue damage. The main objective of PdM is to provide the most precise advance maintenance planning and to avoid unexpected breakdowns. Knowing when a particular machine needs to be serviced makes it easier to plan resources for maintenance work such as spare parts or personnel. In addition, system availability can be increased by converting "unplanned stops" into ever shorter and more frequent "planned stops". Further advantages include potentially longer service life of the plants, increased

A. Bakri and Mohd. A.-F. Mohd Szali Januddi, *Systematic Industrial Maintenance to Boost the Quality Management Programs*, SpringerBriefs in Applied Sciences and Technology, https://doi.org/10.1007/978-3-030-46586-5_3

plant safety, fewer accidents with negative effects on the environment and optimized spare parts handling.

3.1.1 Historical Perspective of CBM

Recently, with the emergence of digital technologies, we can get the visibility of asset status information during its usage period. It gives us new challenging issues for improving the efficiency of asset operations. One issue is to implement the CBM approach that makes a diagnosis of the asset status based on wire or wireless monitored data, predicts the assets abnormality, and executes suitable maintenance actions such as repair and replacement before serious problems happen.

3.1.2 CBM Philosophy

CBM is not a substitute for the more traditional maintenance management methods. It is, however, a valuable addition to a comprehensive, total plant maintenance program. This methodology is often regarded as having existed for many years; it is in fact a recently developed methodology that has evolved over the past three decades from precursor maintenance methods. Where traditional maintenance management program relies on routine servicing of all machinery and fast response to unexpected failures, a CBM program schedules specific maintenance tasks as they are actually required by plant equipment. With the help of PdM technologies, the condition of machines is evaluated in order to predict when maintenance needs to be performed. As a result, cost savings can be achieved over routine or time-based preventive maintenance, as tasks are only performed when they are needed. It cannot totally eliminate the continued need for either or both of the traditional programs, i.e., run-to-failure and preventive; PdM can reduce the number of unexpected failures and provide a more reliable scheduling tool for routine PM tasks.

3.1.3 Operational Approach of CBM

PdM differs from PM in that it is based on the actual condition of a machine and not on its average or expected life to predict when maintenance is required. PdM is based on the theoretical failure rate and therefore disregards the actual machine performance. Downtime is planned on the basis of calendar dates or usage. Care should be taken to ensure that the maintenance time is neither too early nor too late. CBM is a type of PdM that uses sensor devices to collect real-time measurements (i.e., pressure, temperature, or vibration) on a piece of equipment while it is in operation. CBM data allows maintenance personnel to perform maintenance at the exact moment it is needed, prior to failure. The CBM process requires technologies, people skills, and communication to integrate all available equipment condition data, such as diagnostic and performance data; maintenance histories; operator logs; and design data, to make timely decisions about the maintenance requirements of major/critical equipment. CBM monitors the condition or performance of plant equipment through various technologies. The data is collected, analyzed, trended, and used to project equipment failures. With CBM, maintenance is only performed when the data shows that performance is decreasing or a failure is likely. Rather than at specified intervals like with preventive maintenance. CBM uses various process parameters (e.g., pressure, temperature, vibration, flow) and material samples (e.g., oil and air) to monitor conditions. With these parameters and samples, condition-based maintenance obtains indications of system and equipment health, performance, integrity (strength) and provides information for scheduling timely correction action.

3.1.4 Example of Implementation Step for CBM

Predictive maintenance (PdM) is a broad field of many overlapping technologies, all of which have one common goal: find faults in a machine's condition before the machine fails. On their own, predictive technologies each have their own strengths. For example, the following are our most common applications of vibration, infrared, and ultrasound:

- Vibration: diagnose mechanical faults in rotating machines
- Infrared: detect problems in electrical circuits
- Oil analysis: detect deterioration in the lubrication (oil) of the engines.

3.1.4.1 Vibration Monitoring

Vibration Monitoring is a form of condition-based monitoring that involves listening to vibrations in an operating piece of equipment in order to determine whether abnormal vibration patterns exist. Vibration monitoring is an important part of effective asset integrity management, because changes in vibration levels can be indicative of advanced wear and a number of other problems, including equipment coming loose from mountings or malfunctioning parts. Information gathered while monitoring for vibration inconsistencies can be used while planning predictive maintenance activities. One of industries' most fundamental applications is the motor-pump assembly. Many pump problems and failures manifest themselves as vibrations, and monitoring the vibration levels is therefore a profitable strategy toward increasing the operating efficiency, minimizing unplanned downtime, and reducing operation costs (Fig. 3.1).

Typical location to inspect vibration on pump unit.

Fig. 3.1 Location to place the probe for vibration meter

3.1.4.2 Infrared Thermography

Heat is often an early symptom of equipment damage or malfunction, making it a key performance parameter monitored in predictive maintenance (PdM) programs. Infrared thermography is a tool that has become more and more widely used for preventative maintenance on mechanical and electrical systems over time. It takes advantage of the infrared radiation properties to extract useful conclusions for the condition of the equipment under test. It is neither non-destructive, nor an interrupting procedure and has no solid substitute. Technicians who practice infrared predictive maintenance regularly check the temperature of critical equipment, allowing them to track operating conditions over time and quickly identify unusual readings for further inspection. By monitoring equipment performance and scheduling maintenance when needed, these facilities reduce the likelihood of unplanned downtime due to equipment failure, spend less on "reactive" maintenance fees and equipment repair costs, extend the lifespan of machine assets, and further maximize maintenance and production. Here is the trick: To actually save money, predictive maintenance should not create excessive additional maintenance efforts. The goal is to transition maintenance resources away from emergency repairs and into scheduled inspections of key equipment. Inspections take less time than repairs, especially if done with a thermal imager.

A thermal imager takes non-contact, infrared temperature measurements that capture an object's temperature profile as a two-dimensional picture. Unlike an infrared thermometer that only captures temperature at a single point, a thermal

Fig. 3.2 Thermal imaging for identifying electrical issues

imager can capture temperature from both critical components and the entire integrated unit. Thermal imagers can also store previous and current images for comparison and upload images to a central database. Infrared vision and thermography has various implementations in a wide variety of sectors such as medical, military, process monitoring, electrical or mechanical engineering, energy evaluations, R&D. Infrared scanning is recommended as a regular maintenance procedure throughout all industries since no other tool can extract such solid results as quickly and without interrupting the process flow, a benefit essential for the industry regardless of the life "era" of equipment (bathtub model). Initially, baseline data of the equipment needs to be established; this information that could also possibly be useful at a later date in case of warranty claims. In the middle, data to prevent malfunctions is needed before such malfunctions occur by periodic scans that are performed as part of a routine preventive maintenance program, and over time, trends can be evaluated to identify potential problems. It is a great tool to elevate to predictive maintenance. At the end of the life cycle, data to find the exact time line where the non-repairable or uneconomical break down happens is required (Figs. 3.2 and 3.3).

3.1.4.3 Oil Analysis

Oil analysis is another powerful technology which has many applications across different industries. An oil sample can be tested for viscosity, wear particles, the presence of water, and more. In industrial facilities, oil analysis is commonly used to monitor critical equipment such as compressors and gearboxes. In the transportation industry, tests on diesel engines are common and vital. Diesel engines found in trains,

Fig. 3.3 Thermal imaging for identifying electrical issues on pump motor

buses, trucks, and ships are often pushed to the limit in harsh operating conditions. Oil analysis provides an opportunity to monitor the condition of these engines in a way that other predictive technologies cannot.

There are some vital questions that needed to be answered before execute the oil analysis technique. Consider the following: What critical equipment could benefit from oil analysis? What kind of contaminations are the machines exposed to (water, soot, ash, corrosion, etc.)? The answers to these preliminary questions will help determine which tests should be performed on the oil samples and how often samples should be taken. Before sampling begins, make sure to create an equipment list with the machine type, name/asset number, lubricant type/grade/manufacturer, and whether or not the sample is filtered. This information will aid the testing laboratory with analysis. Over time, trends emerge from analyzed oil samples. Sampling regularly will determine optimal lubricant replacement intervals, ensure the machine is operating under proper conditions, and indicate potential warnings of machine failure. In turn, a successful program will increase asset performance and maximize the useful life of those assets (Fig. 3.4).

3.2 Reliability-Centered Maintenance

Reliability-centered maintenance (RCM) is a process used to determine what is the most appropriate maintenance requirement to ensure the assets within organization's

Fig. 3.4 A typical oil analysis for identifying lubrication issues

facility continues its functionality in its present operating context. With regard to the earlier definition of maintenance, a fuller definition of RCM could be "*a process used to determine what must be done to ensure that any physical asset continues to do whatever its users want it to do in its present operating context*" (Moubray). Generally, the RCM technique is applicable to huge, complex, and high risks systems, such as aircraft, oil rig, nuclear, and chemical processing plants.

3.2.1 Historical Perspective of RCM

RCM methodology started in the aviation industry during the 1960s, aimed at optimizing maintenance costs, reducing safety risk, and enhancing the reliability of airplane. By the late 1950s, the cost of maintenance activities in the aviation industry had become high enough to warrant a special investigation into the effectiveness of those activities. Accordingly, in 1960, a task force was formed by US government consisting of representatives of both the airlines and the Federal Aviation Administration (FAA) to investigate the capabilities of existing PM techniques. The establishment of this task force subsequently led to the development of a series of guidelines for airlines and aircraft manufacturers to use, when establishing maintenance schedules for their aircraft. The final outcome of the task force is the final report, written by Stan Nowlan and Howard Heap and published in 1978, was entitled Reliability-Centered Maintenance and has become the report upon which all subsequent RCM approaches have been based. Nowlan and Heap concluded that many types of failures could not be prevented no matter how intensive the maintenance activities. Additionally, it was discovered that for many items the probability of failure did not increase with age. Consequently, a maintenance program based on age will have little, if any effect on the failure rate.

3.2.2 RCM Philosophy

RCM process focuses on preserving an equipment function by applying an appropriate and integrated PM tasks. However, it is different from conventional PM, since its focuses mainly on functionality rather than focusing on equipment in overall. The RCM philosophy employs PM, predictive maintenance (PdM) or condition-based maintenance (CBM), and other proactive maintenance techniques in an integrated manner to increase the probability that an equipment will function as required with a minimum of maintenance. RCM process also embedded other quality and safety improvement tools to enhance its approach. These includes failure mode and effect analysis (FMEA); fault tree analysis (FTA); failure mode effect and criticality analysis (FMECA); and hazard and operability analysis (HAZOP). RCM is a living system—RCM gathers data from the results achieved and feeds this data back to improve design and future maintenance. This feedback is an important part of the proactive maintenance element of the RCM program.

There are four principles that are critical for a reliability-centered maintenance program:

- The primary objective is to preserve system function
- Identify failure modes that can affect the system function
- Prioritize the failure modes
- Select applicable and effective tasks to control the failure modes.

3.2.3 Operational Approach of RCM

There are several ways to conduct and implement an RCM program. RCM analysis is the method first proposed and documented by Nowlan and Heap and later modified by John Moubray, Anthony M. Smith, and others. The RCM process necessitates asking seven questions to identify the functions, the source and effect of the failures about the asset or system under review, as follows:

1. What are the functions and associated desired standards of performance of the asset in its present operating context (functions)?
2. In what ways can it fail to fulfill its functions (functional failures)?
3. What causes each functional failure (failure modes)? In what way? What happens when each failure occurs (failure effects)?
4. In what way does each failure matter (failure consequences)?
5. What should be done to predict or prevent each failure (proactive tasks and task intervals)?
6. What should be done if a suitable proactive task cannot be found (default actions)?

The First Question is about identifying the functions and the performance standards of the asset. Functions of an asset can be categorized into two categories, namely primary and secondary functions. Primary functions are the intended functions in which the asset was designed. For example, the primary function of roller conveyor unit is for transporting of goods. The secondary functions of roller conveyor unit are about its compatibility with other requirement, for safety, health, environment, and compatibility with other equipment in one system. The performance standard can be defined as the ability of an asset to perform according to its design objective.

The Second Question requires user to define the functional failure of an asset. Functional failures describe ways that the equipment may fail to perform its intended function(s) to a standard of performance that is acceptable to the user of the asset. This may include failure to perform a function, poor performance of a function, over-performance of a function, performing an unintended function, etc.

   Failure can be defined in many ways. In a broad sense, failure is simply an unsatisfactory condition. RCM, however, requires us to look at failure from not just an equipment standpoint, but a system standpoint as well. A piece of equipment can be operating (an HVAC unit in a clean room, for example), but if its output is less than required it would be considered failed. A functional failure is essentially the inability of an item/system to meet its specified performance standard. A complete loss of function is a functional failure; however, in the HVAC example above, if the system's output is less than specified, a functional failure has occurred, even if the system is still operating.

The Third Question introduces the significant term in RCM, i.e., failure modes. A failure mode is a cause of failure or one possible way an asset might fail. When an asset has many potential ways of failing, it has multiple failure modes or competing risks. The more complex an asset (for example: a system) is, the more failure modes there are. The failure mode represents the specific cause of the functional failure of an asset. Identifying failure modes is of paramount importance.

The Fourth Question helps user to describe and list what happens when each failure modes occurs. These are known as failure effects. The description of what happens when each failure mode occurs.

The Fifth Question concerns with the description of how the loss of function matters. Failures of an asset are prioritized according to how serious their consequences are, how frequently they occur, and how easily they can be detected. The purpose of the RCM is to take actions to eliminate or reduce failures, starting with the highest-priority ones. Equipment failures may affect safety, operations, and other equipment. The criticality of each of these failure modes can also be considered. The consequences of failure determine the priority of maintenance activities or design improvement required to prevent occurrence. If failure of an item results in little or no consequence, minimal maintenance activities are generally required. If, however, failure of an item results in a large economic hardship, personnel injury, or environmental damage, maximum maintenance activities or a redesign may be called for.

The Sixth Question describes about prevention. It utters the need of proactive measures, performed to predict, prevent, or find failures. With the help of operators, experienced technicians, RCM experts, and equipment experts, the root causes of each of the failure modes can be identified. Root causes for failure of the conveyor could include a lack of lubrication on the rollers, a failure of a bearing, or a loosened belt.

The Seventh Question is backup plan, just in case the prevention measures is not effective enough to avoid the recurring of failure. Any logical measures are applicable, including but not limited to engineering redesigns, and changes/additions to operating procedures or technical manuals.

Answers to these seven questions determine the necessary actions required to maintain the systems or equipment.

3.2.4 Example of Implementation Step for RCM

1. Set up the modular map of the equipment
   Example: Boiler Training Unit (Figs. 3.5 and 3.6)
2. Modular map for Boiler Training Unit
   Example: FMEA/RCM table for Boiler Training Unit
3. Fill up the FMEA/RCM table of the equipment follow the below sequence
   Example: FMEA/RCM table for Boiler Training Unit

   - **Item**: To fill up the part name.
   - **Function**: To fill up the part function.
   - **Potential failure mode**: To identify all failure modes relate to the functional requirements of the equipment part. Examples: deforming, electrical short circuit, corrosion, rupture.
   - **Potential Effect(s) of the failure**: Determine the potential failure effects associated with each failure mode. The effect is related directly to the ability of

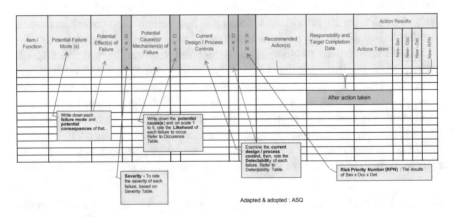

Fig. 3.5 A basic FMEA table

Fig. 3.6 Modular map for boiler training unit

that specific component to perform its intended function. The effect should be stated in terms meaningful to the product or system performance. If the effects are defined in general terms, it will be difficult to identify (and reduce) true potential risks. Examples of failure effects include overheating, noise, abnormal shutdown, user injury. To write down the effects from failure mode.

- **Severity**: To rank the severity of failure. Severity is the seriousness of failure consequences of failure effects. Usual practice rates failure effect severity (Sev) on a scale of 1–5 (or 1–10), in which one is lowest severity and 5 (or 10) is the highest. Table 3.1 shows typical FMEA severity ratings and their descriptions.
- **Potential Cause(s)/Mechanism of Failure**: Examine cause(s) of each failure mode and analyze at similar processes and their documented failure modes. All potential failure causes should be identified and documented in technical terms. Uses of tools classified as root cause analysis (RCA) tool, as well as

Table 3.1 A typical Severity table for RCM

Severity		
Ranking	Severity level	Description/Condition
5	Very high	Equipment inoperable: Loss in primary function (with catastrophic failure)
4	High	Equipment inoperable: Loss in primary function
3	Moderate	Equipment operable: With a reduced level of performance
2	Low	Equipment operable: With a reduced level of performance and affected to product/service
1	Minor	Equipment operable: Almost no effect

the best knowledge and experience of the team, would certainly be benefited. RCA is not a single well-defined method; there are many different processes and methods for performing RCA analysis that are defined by their approach or field of origin. Thus, it is as important to use the right analysis method, so it matches the scope and complexity of the problem. Some of the common RCA techniques are:

Five why—Asks why, why, why until fundamental causes are identified.
Ishikawa/fishbone diagram—Causes and effects are listed in categories.
Cause and effect analysis/Causal factor tree—The causal factors are displayed on a tree so that cause–effect dependencies can be identified.
Fault or logic tree analysis—A failure is identified, and the failure modes are described and tested until the roots are identified.
Pareto chart—Shows the relative frequency of problems or failures in rank order so that process improvement activities can be focused on the significant few.

- **Occurence**: To rank the occurrence of failure based on the Occurrence Table 3.2 as shown in the below example. Examine how often failure occurs.

Table 3.2 A typical Occurrence table for RCM

Occurrence		
Ranking	Frequency	Description/Condition
5	Very high	Failure is almost inevitable (ex.: 1 in 2 weeks)
	High	High repetitive of failures occurred in certain period of time (ex.: 1 in 3 months)
3	Moderate	Occasional failures occurred in certain period of time (ex. 1 in 5 months)
2	Low	Relatively a few failure occurred in certain period of time (ex. 1 in 8 months)
1	Remote	Failure is unlikely to occur in certain period of time (ex. 1 in 12 months)

Table 3.3 A typical Detection table for RCM

Detection		
Ranking	Detection level	Description/Condition
5	Almost impossible	Almost impossible to detect failure mode during equipment in operation
4	Difficult to detect	Failure mode can be detected during random check/routine maintenance
3	Moderate	Failure mode detected through emergency light/buzzer during equipment in operation
2	Easy to detect	Failure mode detected by automated controls that it will detect discrepant part and automatically prompt error
1	Very easy to detect	Failure mode detected at its source

Data can be obtained from maintenance record or any relevant documentation. Usual practice rates for occurrence (Occ) is based on a scale of 1–5 (or 1–10), in which one is lowest (remote) and 5 (or 10) is the highest. The following table shows typical FMEA occurrence ratings and their descriptions.

- **Current Design/Process control**: All potential failure causes should be identified and documented in technical terms. Failure causes are often indicative of weaknesses in the design or process control. Examples of causes include poor material used in design, unsuitable operating environment, lack of enforcement, insufficient skills.
- **Detection**: To rank the detection of failure based on the Detection Table 3.3 as shown in below example.
- **RPN**: The RPN is calculated by multiplying the three scoring columns: severity, occurrence, and detection. For example, if the severity score is 4, the occurrence score is 3, and detection is 3, then the RPN would be 36.
- **Recommended action(s)**: The action plan outlines what steps are needed to implement the solution, who will do them, and when they will be completed. Most action plans identified during a PFMEA will be of the simple "who, what, and when" category. Responsibilities and target completion dates for specific actions to be taken are identified. Taking action means reducing the RPN. The RPN can be reduced by lowering any of the three rankings (severity, occurrence, or detection) individually or in combination with one another. After remedial actions are determined, they should be tested for effectiveness and efficiency.
- Action Results: When corrective measures are implemented, RPN is calculated again and the results are documented in the FMEA (Table 3.4).

Table 3.4 Example of RCM - based on FMEA table

Machine name	Boiler training unit		Main PIC: Adnan Bakri
Model	XYX12345		Prepared by: Fatihhi Januddi
Core team	Adnan Bakri, Fatihhi Januddi, Shahrul Affendy Kosnan, Zulhaimi Mohamad		

No.	(1) Item	(2) Function	(3) Potential failure mode(s)	(4) Potential effect(s) of the failure	Severity (5)	(6) Potential cause(s)/mechanism of failure	Occurrence (7)	(8) Current design/Process control	Detection (9)
FMEA/RCM Serial no: PQRS3564	Pressure gauge	Used to indicate the steam pressure of the boiler	Intermittent	Pressure reading inaccurate	4	The calibration schedule was not followed	3	Calibration schedule to check the pressure gauge was set to every 6 months	3

FMEA/RCM Serial no: PQRS3564

Checked by: Mohd Zul-Waqar Bin Mohd Tohid

Approved by: Mohamad Bin Najib

RPN (10)	(11) Recommended action(s)	(12) Action results				
		PIC	New—Sev	New—Occ	New—Det	New RPN
36	Enforcement to ensure the technician follow the calibration schedule					

References

1. General Information (1) on Predictive Maintenance (2019) Available at: https://www.advanc
   edtech.com/wp-content/uploads/2017/11/PDM-Condition-Monitoring.pdf. Retrieved date: 13
   February 2019
2. General Information (2) on Predictive Maintenance (2019) Available at: https://www.fiixsoftw
   are.com/maintenance-strategies/predictive-maintenance/. Retrieved date: 20 April 2019
3. General Information (3) on Predictive Maintenance (2019) Available at: https://www.reliablep
   lant.com/Read/12495/preventive-predictive-maintenance. Retrieved date: 20 April 2019
4. General Information (1) on Reliability Centered Maintenance (2019) Available at: https://ins
   pectioneering.com/tag/reliability+centered+maintenance. Retrieved date: 13 February 2019
5. General Information (2) on Reliability Centered maintenance (2019) Available at: https://www.
   weibull.com/basics/rcm.htm. Retrieved date: 13 February 2019
6. General Information (3) on Reliability Centered Maintenance (2019) Available at: https://fas.
   org/sgp/crs/misc/R44776.pdf. Retrieved date: 13 February 2019
7. Moubray J (1997) Reliability-centered maintenance second edition. Industrial Press. ISBN-10:
   0831131462, ISBN-13: 978-0831131463
8. Susto GA, Schirru A, Pampuri S, McLoone S, Beghi A (2015) Machine learning for predictive
   maintenance: A multiple classifiers approach. IEEE Transactions on Industrial Informatics,
   https://doi.org/10.1109/TII.2014.2349359 11(3):812–820
9. Bakri A, Rahim AR, Yusof M (2014) Maintenance management: Rationale of TPM as the
   research focus. In: Applied Mechanics and Materials, Vol. 670. Trans Tech Publications Ltd, pp
   1575–1582

Chapter 4
Integration of Maintenance Activity with QI initiatives

4.1 Total Productive Maintenance

Total productive maintenance (TPM) is a comprehensive maintenance management approach aimed to improve the overall effectiveness of the equipment. It integrates the role of maintenance and production department of a company. TPM represents a radical shift from the conventional view about maintenance. It is considered as a strategic initiative for improving quality by focusing on maintenance activities. TPM is also viewed as an application of TQM approach to equipment, focusing on minimizing production loss.

4.1.1 Historical Perspective of TPM

TPM was first developed in 1969 at Nippondenso, one of the companies in Toyota supply chain. The early development of TPM in Japanese automotive industry was under the leadership of Seiichi Nakajima, the head of the Japan Institute of Plant Maintenance (JIPM). Since then, TPM methodology has evolved and extended to other manufacturing sectors worldwide. Seiichi Nakajima was recognized as the "father of TPM" for his utmost contribution to support hundreds of plants worldwide in implementing TPM methodology.

4.1.2 TPM Philosophy

Philosophically, TPM shares commonality features with TQM philosophy, particularly in employee participation, cross-functional training, empowerment of employees through a small group activity (SGA), focus on plant efficiency through quality, and emphasize on continual improvement. The entire philosophy of TPM

A. Bakri and Mohd. A.-F. Mohd Szali Januddi, *Systematic Industrial Maintenance to Boost the Quality Management Programs*, SpringerBriefs in Applied Sciences and Technology, https://doi.org/10.1007/978-3-030-46586-5_4

Fig. 4.1 The meaning of total in TPM. Adopted from Nakajima (1988)

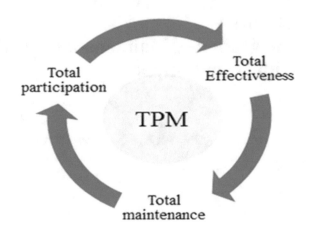

as productive maintenance aims to maximize OEE through a total participation of all level of employees in the operational hierarchy. Under TQM philosophy, defects of the products are eliminated at their processes rather than scrutinized only at the finished product. TPM aims toward achieving zero equipment breakdowns. It is a concept adopted from TQM approach toward zero manufacturing defects and minimizing production losses. The definition of equipment effectiveness is no longer restricted to availability. However, it encompasses quality as the complementary factor. The word "total" in TPM as defined by Nakajima (1988) has three meanings as depicted in Fig. 4.1. It outlines three principal features of TPM. Total effectiveness describes TPM aims toward economic efficiency particularly in maximizing productivity (P) without compromising on quality (Q), optimizing operational cost (C), timely delivery to customers (D), improving the safety, health, and environment of the workplace (S), and boosting the morale of employees (M). On the other hand, total maintenance describes TPM aims toward a systematic maintenance of equipment through a continual maintenance improvement activity. Total participation denotes the need of holistic involvement of all employees.

4.1.3 Focus of TPM

TPM brought out maintenance as an essential and vital part of organizational activities toward business survival. It involves a mutual relationship among all level of employee within operational hierarchy to enhance the OEE. The three ultimate focuses of TPM are to achieve zero breakdowns, zero defects, and zero industrial accidents from operation processes. The essence of TPM is to promote an autonomous maintenance (AM) culture in which operators taking care of the equipment by carrying out PrM work together with maintenance personnel. TPM has been described as a strategic maintenance activity comprising of five elements as listed below:

1. Maximization of the equipment effectiveness by improving the equipment availability and efficiency.
2. Establishment of a comprehensive planned maintenance activity covering the entire life span of equipment.
3. Implementation by cross-functional departments (engineering, production, maintenance, etc.).
4. Involvement of all employees in operational hierarchy.
5. Activation of small group activities, such as AM activities.

4.1.4 Benefits of TPM

Successful TPM implementation would facilitate in achieving numerous organizational priorities and goals as shown in Table 4.1. On top of that, TPM implementation in an organization would also results in intangible benefits such as improved image of the organization in the business. Such intangible benefit would create an opportunity of increasing the order value. Once introduced to the AM activity, the sense of operators' ownership to their equipment will improve; they would be willing to take care of the equipment themselves without being instructed. With the achievement of zero breakdown, zero defects, zero accidents, and zero waste from operation processes, the operators would be motivated to improve their abilities and appreciate the significance of employee contributions toward realization organization goals and objectives.

Table 4.1 Priorities and goals of TPM

Manufacturing priorities	TPM benefits
Productivity (P)	Improve equipment availability and productivity through reduction in unplanned stoppages and breakdown
Quality (Q)	Reduce quality issues from problematic production equipment Eliminated process failures through improvement activity
Cost (C)	Efficient maintenance management and techniques Reduce waste from production processes
Delivery (D)	Support JIT initiatives through reliable equipment
Safety (S)	Improve workplace environment by eliminating hazards
Morale (M)	Inculcate continual improvement culture Increase employees' knowledge on process, product and problem-solving Promote employee involvement and empowerment

4.1.5   Comparison of Various Approaches in Maintenance Management

Table 4.2 summarizes various approaches in maintenance management.

Among presented in Table 4.2, BM is the oldest while TPM is the contemporary maintenance approach. TPM does not fully contradict the rationale of BM; however, its goal is focused on a holistic approach. TPM also integrates PM, and PdM activities as well aimed at maximizing the OEE.

In TPM, PM is directed toward PdM, aimed to detect more effectively any deterioration and failure on equipment through application of CBM technology such as spectroscopy, thermography, and others. Notably, TPM encompasses all elements applied by other maintenance methodology, from tools and techniques to involvement of all operational hierarchy in the organization. What is very much required in manufacturing organization is to integrate different functional areas in a coherent manner.

The integration of maintenance and quality improvement initiatives, particularly in TPM and TQM, is the upshot of the need of strategic manufacturing initiatives toward a superb quality and competitive products. The pressure to compete in domestic as well as global market has forced manufacturers to embark on the quality initiatives and to focus more on equipment performance. Due to that, the view on the maintenance role was tremendously changed from as a cost generator to as one of the main contributors to business profitability. TQM and TPM share a lot of commonalities such as total employee involvement and continual improvement.

4.1.6   Operational Approach of TPM

The foundation for operational approach of TPM is called pillars of TPM. The definition of the pillars varied and tailored according to organizational goals and objectives. Figure 4.2 shows TPM model proposed by JIPM which consists of eight pillars. It has been observed that many organizations strictly followed the JIPM-TPM model in implementing TPM. However, some of TPM consultants and practitioners in the West have simplified the JIPM-TPM model by eliminating and integrating some of the pillars. For instance, Yeomans and Millington (1997) revised the JIPM-TPM model and developed a new model consisting of five pillars, which are increased improvement effectiveness, training, autonomous maintenance, early equipment management, and preventive maintenance. Steinbacher and Steinbacher (1993) also revised the conventional JIPM-TPM model by integrating training and education pillar with other pillars. Figure 4.3 shows the revised TPM pillars by Yeomans and Millingston (1997) and Steinbacher and Steinbacher (1993).

Table 4.2 Comparison on the approaches in maintenance management

Descriptions			Maintenance management					
			BM/CM	PrM	PdM/CBM	RCM	CMMS	TPM
1	Type of maintenance		Reactive	Proactive	Proactive	Proactive	Proactive	Proactive
2	Effectiveness in managing maintenance of equipment and system		X	✓	✓	✓	✓	✓
3	Application tools and techniques	Common tools and techniques	✓	✓	✓	✓	✓	✓
		Advanced repair tools and techniques	X	X	✓	✓	X	✓
4	Operational involvement	Maintenance staff	✓	✓	✓	✓	✓	✓
		Production staff	X	X	X	X	X	✓
		Other staffs/departments	X	X	X	X	X	✓
5	Integration with other initiatives in organization	Quality initiatives (TQM, JIT, lean)	X	✓	✓	✓	✓	✓
		Safety and environmental management	X	✓	✓	✓	✓	✓
		Small group activities	X	X	X	X	X	✓

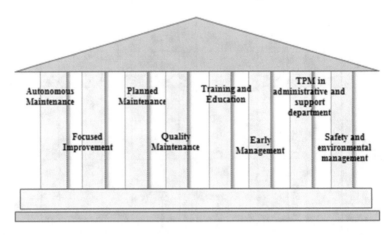

Fig. 4.2 TPM with eight pillars approach. *Note* Adopted from Wakjira and Singh (2012), Gupta and Garg (2012), Ahuja and Kumar (2009), Ahmed et al. (2005), Ireland and Dale (2001), Tsang and Chan (2000), Suzuki (1994)

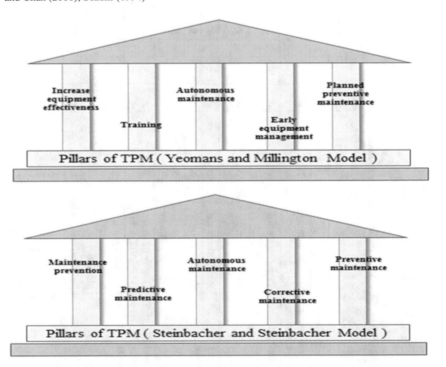

Fig. 4.3 TPM with five pillars approach. Note Adopted from Yeomans and Millington (1997), Steinbacher and Steinbacher (1993)

Table 4.3 Seven steps AM

AM step	Activity	Description
Step 1	Perform initial cleaning	Check equipment and expose irregularities (*Denoted as C*)
Step 2	Control contamination sources and inaccessible place	Act against contamination sources and inaccessible places (*Denoted as A*)
Step 3	Establish cleaning and checking standards	Plan and do checks based on standards (*Denoted as P, D*)
Step 4	Perform general equipment inspection	Repeat the $C \to A \to P \to D$ cycle for each category
Step 5	General process inspection	Repeat the $C \to A \to P \to D$ cycle for each category
Step 6	Systematize autonomous maintenance.	$C \to A \to P \to D \to C \to A \to P \to D$
Step 7	Practice full self-management	$C \to A \to P \to D \to C \to A \to P \to D$

Adopted from Suzuki (1994)

4.1.6.1   TPM Pillar 1: Autonomous Maintenance

Autonomous maintenance (AM) is described as maintenance performed by production department, and it is one of the basic structures in any TPM activity. AM comprises of any maintenance activity performed by the production department intended to ensure equipment operated efficiently to meet the production plan. AM approach will gradually make a shift in thinking of equipment operators. The operator will start to think about their responsibility and capability. The basic knowledge and skills training in AM such as machine cleaning, lubricating, tightening, and minor repair (CLTM) will enhance operator capability and indirectly promote the sense of ownership for their equipment. The goals of AM program are: (1) to avoid the deterioration of equipment through proper operation and maintenance activity; (2) to restore equipment to its optimum condition through systematic maintenance management; and (3) to create basic procedures for minor equipment maintenance. AM is implemented in seven-step approach starting with initial cleaning. It promotes the establishment of optimal process conditions through the Check–Act–Plan–Do (CAPD) continuous improvement cycle shown in Table 4.3.

4.1.6.2   TPM Pillar 2: Focused Improvement

Focused improvement (FI) is activity targeted to optimize the overall effectiveness of equipment through elimination of all kinds of losses. Identifying and quantifying those losses are therefore important issues. The traditional method of identifying losses analyzes results statistically to identify problems, then studies to find their causes. The method adopted in TPM emphasizes on a hands-on, practical approach and examines all production inputs particularly on equipment (machine), materials, people (man), methods and environment. It analyzes those inputs thoroughly and treats any deficiencies in these inputs as losses. FI embedded Kaizen methodology in its approach. The word Kaizen is a Japanese word for continual improvement. The word "Kai" means change, while the word "Zen" means good for the better. FI pillar aimed to reduce the losses on the equipment that affect its efficiency by applying various Kaizen tools. The common tools used in Kaizen activity include why-why analysis, phenomenon-mechanism (P-M) analysis, cause-and-effect analysis (4M1E), fault tree analysis (FTA), and failure mode and effect analysis (FMEA).

4.1.6.3 TPM Pillar 3: Planned Maintenance

In TPM development program, scheduled or planned maintenance (PM), is the deliberate, methodical activity of establishing and maintaining optimal equipment condition. The PM activity would benefit the company toward an efficient and cost-effective maintenance management. PM aims to optimize the equipment with zero failures and zero defects at minimum cost. The contributions of AM and PM enable the production operator to run the equipment effectively and preventing deterioration. PM embraces activities that improve equipment, as well as activities that improve maintenance technology and skill. PM is performed based on two stages: the first stage is carried out through AM activity by the production department, while the second stage focuses on specified maintenance work by the maintenance department. Figure 4.4 shows the relationship between the two main activities in PM.

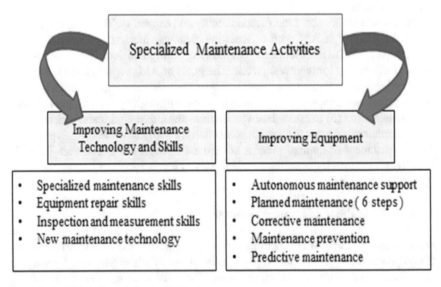

Fig. 4.4 Activities of specialized maintenance

An efficient PM program consist of other maintenance practices and approaches like PM and PdM. PM and PdM are combined together to enhance the control and maintenance of equipment components. Maintenance activities such as inspection, servicing, parts replacement, and cleaning of equipment in PM would be done based on specific time (periodically). It is aimed at preventing sudden failures and process problems. PdM uses equipment diagnostics techniques to monitor and diagnose the condition of rotating or moving equipment.

4.1.6.4 TPM Pillar 4: Quality Maintenance

Quality maintenance (QM) aimed at satisfying customer through defect-free production. It focuses on eliminating non-conformances in the production processes in a structured approach. QM brings a new concept in controlling quality by promoting reactive to proactive approach, that is from quality control to quality assurance. QM evolved from the concept of perfect equipment will produce perfect or quality products. The equipment condition is inspected and measured in specific time to ensure the measured values are within the standard specification. The fluctuation of measured values is observed and analyzed. From such activity, the possibilities of defects occurring can be predicted and the appropriate countermeasures can be executed beforehand.

4.1.6.5 TPM Pillar 5: Training and Education

The goal of training and education (TE) pillar is to produce multi-skilled employees. The focus is on achieving zero losses caused by sufficient knowledge, skills, and techniques of work. A continuous TE will improve those deficiencies among the entire workforce in the organization. TE for operators is a fundamental tool of TPM that is always ignored by some organization. The skill development of equipment operators is crucial since they are the one close and have a specific knowledge on the equipment; therefore, they should be included in all phases of TPM. The operator should be equipped with both theoretical and technical knowledge about the equipment, to enable them to do proper maintenance of the equipment. Well-trained operator will understand the mechanism of failures on equipment. They will be able to provide useful suggestion to prevent the recurrent of failures.

4.1.6.6 TPM Pillar 6: Early Management

Equipment management (EM) aimed to ensure the equipment performs at optimum level throughout its entire life span. In process industries, major equipment items are often customized to individual specifications. They are sometimes designed, fabricated, and installed in a rush. Without strict early management, such equipment enters the test operation phase with many hidden defects. The truth of this is borne out by frequency with which maintenance and production personnel discover defects generated in design, fabrication, and installation during shutdown maintenance and start-up. EM is particularly important in process plants because large amount of money is invested in their processing units and management expects the plant to operate for a considerable number of years. EM should be emphasized as equally important as the other TPM activities. The basis of EM is for performance evaluation (optimizing life-cycle costs). The life-cycle cost of equipment item is total cost over its whole life. EM activity will minimize the upcoming maintenance costs. EM activity improves the design of the new equipment by considering the existing maintenance history on the current equipment. Neglecting EM on equipment will inflate operating costs and impairs operability and maintainability during commissioning, test, and shutdown maintenance.

4.1.6.7 TPM Pillar 7: TPM in Administrative and Support Department

TPM in administrative and support department (TPMAS) is also known as office TPM. It focuses to identify and eliminate losses from administrative and production support department. The existing processes and procedures of administrative and production support department will be analyzed thoroughly in order to raise its efficiency. Unlike production department, the supporting department such as planning, engineering, quality control, facilities, and administration do not add value directly. As experts in their particular area, the primary responsibility of those departments is to process information, advice, and assist the production department to reduce costs. Information from departments such as engineering and administration triggers action in the production department. The quality, accuracy, and timing of that information, therefore, profoundly affect what the production department does. How this information is handled is the main concern of TPMAS. In TPM, the work in such departments is treated as analogous to a production process, with the administrative procedures viewed as counterparts of production equipment. The soft-approach of administrative function focuses on improving work allocation and administrative procedures, while the hard-approach improves office layouts and equipment such as computers, copiers, desks, lockers. Tangibly, those arrangements will create a neat office environment and intangibly, and it will result in increased employee efficiency psychologically and physically. From this perspective, the AM steps and shop floor control (5S) principles implemented on the plant floor area easily translated to the office environment.

4.1.6.8 TPM Pillar 8: Safety, Health and Environment

The basic principles of TPM targeted to ensure the reliability of equipment, avoiding human fault and eliminating accidents. Therefore, safety, health, and environmental management (SHEM) is one of the key activities in TPM implementation program.

A full TPM implementation will improve safety, health, and environmental management in many ways, for examples: reconditioned the damaged equipment parts which is one type of a common hazard in the workplace; a thorough application of 5S principles (as part of AM) eliminates leaks and spills and makes workplaces clean, tidy, and well-organized; an active AM and FI activities will eliminate the unsafe areas.

4.1.6.9 Overall Equipment Effectiveness

The performance of TPM is measured based on the performance metric called overall equipment effectiveness (OEE). OEE is interpreted as the multiplication of availability, performance efficiency and rate of quality as shown below:

$$OEE = \text{Availability } (A) \times \text{Performance Efficiency } (P) \times \text{Rate of Quality } (Q)$$

1. Availability $(A) = \frac{\text{Loading time} - \text{Downtime}}{\text{Loading time}} \times 100$
2. Performance efficiency $(P) = \frac{\text{Processed amount}}{\text{Operating time/Cycle time}} \times 100$
3. Rate of quality $(Q) = \frac{\text{Processed amount} - \text{Defect amount}}{\text{Processed amount}} \times 100$

World-class equipment performance is evaluated based on the achievement of 85 per cent of OEE. That formula was resulted from the overall achievements of 90% of equipment availability, 95% of performance efficiency, and 99% of quality rate. The typical OEE value found in an average manufacturing are 35–45%. Plants that have made good productivity gains may operate with OEE values of 50–70%. To optimize the OEE, TPM focuses on reducing and eliminating the six major losses in operation; these include losses due to equipment breakdown; losses from setup and adjustment; losses due to idle and minor stoppages; losses due to reduction in speed; losses due to defect and rework; and losses due to poor yield. The calculation of OEE by considering the impact of those six major losses is shown in the Fig. 4.5, and then, further elaboration of these losses are tabulated in Table 4.4.

4.1.7 Example of Implementation Steps for TPM

It takes at least three to five years for the company to achieve the results from TPM implementation. Table 4.5 shows the three basic stages in TPM development to be followed by any company in order to have a success implementation. The stages involved are preparation stage, implementation stage, and stabilization stage.

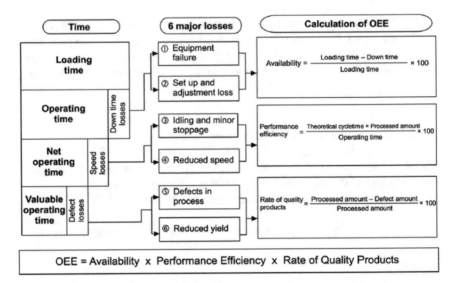

Fig. 4.5 Six major losses related equipment maintenance

Table 4.4 Detail of six major losses related to equipment maintenance

Type of losses	Causes
Breakdown losses	This loss includes equipment downtime, man power, and spare parts cost
Setup and adjustment losses	This type of losses caused by changes in operating conditions, such as the commencement of production runs, start-up at each shift, changes in products, and conditions of operation. These losses consist of downtime, set up (equipment changeovers, exchanges of dies, jigs, and tools), start up, and adjustment
Minor stoppage losses	Caused by events such as machine halting, jamming, and idling
Speed losses	Caused by reduction in operating speed in which the machine cannot be operated at original or theoretical speed
Quality defect and rework losses	Caused by off-specification or defective products manufactured during normal operation. These products must be reworked and scrapped. The losses consist of the labor required to rework the products and the cost of material to be scrapped
Yield losses	Caused by unused or wasted raw materials and are exemplified by the quantity of rejects, scraps chip, etc. The yield losses are divided into two groups. One is the raw material losses resulting from product designs, manufacturing methods, and equipment restrictions. The other group is the adjustment losses resulting from quality defects associated with stabilizing operating conditions at the commencement of work (start-up), changeover, etc.

Table 4.5 Twelve-step TPM implementation

Stage	TPM implementation steps	Activities involved
Preparation	Announcement by top management on the company's decision to introduce TPM	Declare in TPM in-house seminar
	Execution of promotional and training activity	Published in organization bulletin
	Establishment of steering committee	Managers: trained in seminar/camp at each level General employees: seminar, meetings using slides Formation of committees and sub-committees
	Establishment of TPM policies and goals	Benchmarks and targets developed Prediction of effects
	Establishment of TPM master plan	Develop step-by-step TPM implementation plan Framework of strategies to be adopted over time
Implementation	Kickoff ceremony	Invite suppliers, related companies, and affiliated companies
	Improvement of OEE	Pursuit of improvement of efficiency in production department Project team activities and SGA
	Development of AM	Step system, diagnosis, qualification certification
	Development of PM management for the maintenance department	Improvement maintenance, periodic maintenance, predictive maintenance
	Development of TE to improve operation and maintenance skills	Group education of leaders and training members
	Development of EM program Establish QM Establish TPMAS to improve efficiency of administration and other indirect departments Establish SHEM to establish systems to control safety, health and environment	Development of easy to manufacture products and easy to operate production equipment Setting conditions without defectives, and its maintenance and control Support for production, improving efficiency of related sectors Creation of systems for zero accidents and zero pollution cases
Stabilization	Completion of TPM implementation	Sustaining maintenance improvement efforts Challenging higher targets Applying for TPM awards

4.2   Autonomous Maintenance

As discussed in previous chapter, total productive maintenance (TPM) is a comprehensive maintenance management approach aimed to improve overall effectiveness of the equipment. It integrates the role of maintenance and production department of a company. One of the crucial components (pillars) of TPM is autonomous maintenance (AM).

4.2.1   AM Philosophy

AM is "independent" maintenance carried out by the operators of the machines rather than by dedicated maintenance technicians. It is core concept of TPM. AM gives more responsibility and authority to the operators, as the people that have day-to-day contact with the machines and are the most familiar with the operation of the machine. This approach enables them to feel greater ownership for their work and become more involvement in the activities related to the machine such as operation, maintenance, and improvement. By doing so, it would release the technical personnel to do more specialize tasks such as preventive and improvement works of the machine. AM approach is contrast to the traditional maintenance approach, in which the operators run the machines until they break and then handed over to the maintenance personnel for repair works. In such situation, the operator (production) and maintenance have become detached entities and the "I make, you fix" mentality has pervaded the shop floor, in which the operators feel themselves simply as 'producers," whose only contact with the equipment during operation only quality check and any related works to maintain the equipment, even lubricating it and keeping it clean, is seen as the jurisdiction of a maintenance team. In AM approach, the operators will perform the minor (simpler) maintenance routines such as cleaning–lubrication–tightening–minor repair (CLTM). Therefore, it is crucial for operators to have full training and understand the functioning of the equipment.

4.2.2   Operational Approach of AM

The foundation for operational approach of AM is called as "Step of AM." Fig. x shows the step of AM proposed by JIPM which consists of seven steps. The seven steps of AM follow three phases. In the real practice of AM, there is an additional step, identified as "Step 0" which aimed at equipping the operator knowledge on the equipment functions and safety risks. The seven steps for developing AM can be grouped into the three phases described in Table 4.6.

If it is going to work properly, AM must be implemented in accordance with a management-led plan. The AM program should be implemented with each step

Table 4.6 Autonomous Maintenance steps

Phase	AM steps
Preparation Phase	Step 0
Phase-I	Step 1: Initial cleaning
	Step 2: Eliminate sources of contamination and hard-to-access areas
	Step 3: Provisional AM standard (provisional of CLTM standards)
Phase-II	Step 4: General inspection
	Step 5: Autonomous inspection
Phase-III	Step 6: Standardization
	Step 7: Full AM management

being reviewed by the management before authorization is given to proceed to the next.

4.2.2.1 AM Step 0

The goal of Step "0" is to improve the fundamental knowledge about machine components as well identifying its safety risks (hazard). The next goal is to expose and train machine operators to perform the basic maintenance activities, i.e., CLTM. A structured training module is crucial to train machine operators to perform such basic maintenance tasks effectively. A proper training program can standardize the tasks and assists machine operators to comply with the correct standards. Technical experts within the company are good assets that can be utilized to conduct training programs for the operators.

4.2.2.2 Phase I

Phase I constitutes Steps 1–3. The key activities in this phase are about cleaning and inspection. Every single parts of the machine should be thoroughly cleaned and inspected. The team involved in AM are driven toward understanding this important process that will transform the machine conditions to as much as new condition (refurbished machine). Everyone in the AM team will be exposed to the significant function of cleaning and inspection, like the following:

- 'Cleaning is inspection'—Cleaning done is not merely be cleaning for its own sake; however, cleaning is done with intention to find problem.
- 'Inspection is detection'—A thorough inspection throughout cleaning will reveal any kinds of problem (imperfections) of the machine.

Any problem or imperfections detected during cleaning and inspection exercise would provide a room for making further improvements. During this phase, everyone in AM team needs to get into the habit of exercising creativity in solving problems. The CLTM carried out at this phase in which the basic machine conditions are sustained, are the minimum prerequisites for preventing machine deterioration, and established the foundation for continual improvement of the machine.

4.2.2.3   Phase II

Phase II constitutes Steps 4 and 5. In this phase, the AM team will be equipped with further inspection skills and may start performing these inspections themselves (autonomous inspection). This enables them to move on from preventing deterioration to measuring deterioration. At the end of this phase, they should be able to measure (gauge) the level of problem (imperfections) rather than merely do the prevention activity only. Through this, the AM team become truly machine-competent, capable of performing routine inspection using their five senses backed up by good judgment.

4.2.2.4   Phase III

Phase III consists of Steps 6 and 7. This is the final phase that will drive the AM team toward initiating improvements on their own initiative. This is the phase at which the results began to visualize and people's attitudes change. These new attitudes produce a new atmosphere, as operators start to have sense of ownership and have pride in taking care of their machine. At the same time, the results will be visualized in terms of defect reduction and decreasing breakdowns. This is a crucial part of creating a real autonomous management on machine and workplace.

4.2.3   Example of Implementation Steps for AM

4.2.3.1   Step 0: Preparation Phase

Set up the modular map of the machine.
   Example: Lathe Machine Training Unit.
   To draw simple diagrams illustrating the parts/mechanisms used by the machine. AM team need to know the names of the parts and understand how their machine works and what kinds of problems happen when it gets dirty and unattended (Fig. 4.6).

4.2.3.2   Set up the Machine Modular Map and Hazard List

The unsafe condition and actions that could occur as during carrying out Step 1 (initial cleaning) need to be listed, and a preventive strategy should be formulated

Fig. 4.6 Modular map for Lathe Machine Training Unit

for each. Thorough safety training should be provided by Safety Officer to avoid any injuries and mishaps (Table 4.7).

4.2.3.3 Step 1: Initial Cleaning

- Before start the initial cleaning, the machine should be shut down (power off).
- AM teams consisting of production, maintenance, and engineering staff may start to lock off the machine and then perform an in-depth cleaning and inspection by looking for any signs of deterioration.
- Any parts that need replacement, refurbishment, and modification must have addressed accordingly.
- Once initial cleaning completed, all the machine covers can be fixed back.
- In some cases, the old machine cover can be painted up with new paint as to make it more shining as new.

4.2.3.4 Step 2: Eliminate Sources of Contamination and Hard-to-Access Areas

- Once machine has been restored, the AM team need to ensure it does not deteriorate again by controlling all the contamination from it source.
- Any problem and inaccessibility to areas that are hard for CLTM activities must be improved.

Table 4.7 Example of machine part name and hazard list

No.	Part name	Descriptions	Hazard	Preventive measure
1	Chuck	Chuck is used to hold the workspace It is bolted on the spindle which rotates the chuck and work piece		
2	Tool post	It is bolted on the carriage It is used to hold the tool at correct position. Tool holder mounted on it		
3	Tail stock	Tail stock situated on bed It is placed at right hand side of the bed The main function of tail stock to support the job when required. It is also used to perform drilling operation		
4	Speed controller	Speed controller switch is situated on head stock which controls the speed of spindle		
5	Lead screw	Lead screw is situated at the bottom side of bed which is used to move the carriage automatically during thread cutting		

- The team need to think creatively in searching for the root causes of contamination (sources of dirt, leaks, and so on).
- Activities in Step 2 is a crucial process that nurtures the seeds of improvement, as AM team will start find ways to improve the situation on their own initiative.
- It allows them to derive real pleasure from the process of improvement and the results attained, and to share a sense of achievement with all team members and management.

4.2.3.5 Step 3: Provisional AM Standard (Provisional of CLTM Tandards)

- Step 3 establish provisional of CLTM standard, in which these standards are the most visible evidence of AM activities.
- Follow the standard procedure of CLTM, the AM team will follow the lubrication and inspection schedule, noting any problems with accessibility, lubrication flow, etc., and develop their own standard indicating items to be cleaned, checked, or lubricated, the methods to be used and frequency and responsibilities.
- However, some technical checks may still be the responsibility of maintenance rather than production, and this is noted in the documentation.

- Step 3 also involves the visual management of the machine and inspection process—marking gauges and sight glasses and even visually numbering the inspection route to prevent checks being missed.
- The final outcome of Steps 1–3 should be a restored and improved machine, with a visually standard for CLTM.
- Step 3 is an important step in which the AM team use the experience they have acquired in Steps 1 and 2 to clarify what the ideal conditions for their machine should be, and devise standards for the actions necessary to sustain those conditions (standards specifying the 5 Ws and 1 H, i.e., who is to do what, where, when, why, and how).
- The AM audit system assesses the progress made by AM teams toward achieving the company's TPM goals. AM audits determining whether or not the teams are ready to move up to the next level of maturity with respect to the seven steps of the AM pillar.
- Figure 4.7 shows a typical template for creating a provisional standard, and Fig. x shows an example of audit form used to gauge the AM (Steps 1–3) implementation.

The consequences step, i.e., Steps 4–7, is aimed achieving two main objective of AM, i.e., restoring machine condition and developing the small group activities (developing team autonomy). The both objective will ensure the production team develop ownership of their machine, learn to set their own goals in line with company policy, and manage their own improvement activities (Figs. 4.8 and 4.9).

Cleaning-Inspection-Lubrication-Minor Repair/Tightening (CLTM) Daily Standard

Dept	: A&T	Area	: Warehouse A2	Date	: 26/12/2019	Approval		
Section	: FAME	Machine	: Lathe Machine -AP2020	HOD	: Abu Zar			

No	Diagram	Part name	Type of CLTM & Criterion		Tools	Time (min)	Frequency				PIC
							S	D	W	M	
1		Chuck	C	Clean from dirt, metal chips	Cotton rage, scrapper	2	↗				
			L								
			T								
			M								
2		Tool post	C	Clean from dirt, metal chips	Cotton rage, scrapper	2	↗				
			L								
			T								
			M								
3		Tail stock	C	Clean from dirt, metal chips	Cotton rage, scrapper	2	↗				
			L	Sufficient lubrication.	Lubrication oil - NEG 14	4	↗				
			T								
			M								

Fig. 4.7 A typical AM/CLTM checklist

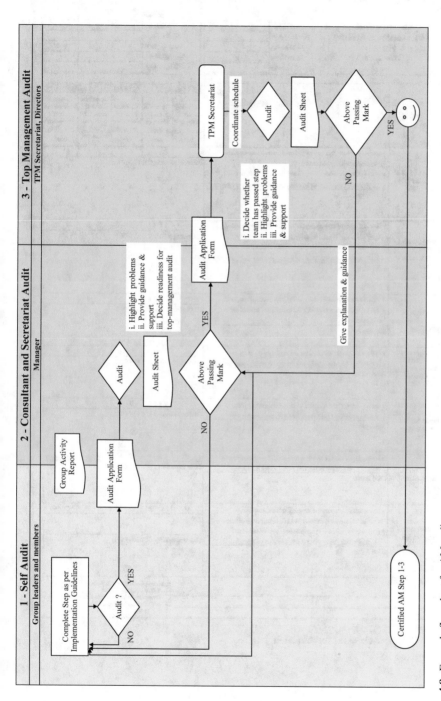

Fig. 4.8 Example flow chart for AM audit

Autonomous Maintenance Audit Form STEP 1,2,3			Self Audit	D e pa r t	Top Mgmt
	Department	Audit Date			
	Process / Area	Audit No.			
	Machine No.	Auditor Name			
	Team Name	Passing Marks	80%	70%	60%
	Leader	Audit Points			
		Auditor Sign.			

Note: Circle 'O' your score

Items	Description	Almost completely not implemented (Bad)	Implemented insofar as can be seen (Poor)	Implemented as far as specified locations (Fair)	Implemented even for parts that cannot be seen (Somewhat Good)	Thoroughly implemented, and proceeding with HTAA (Good)
1. Cleanliness of machine's main body	▸ How well is visual floor control (indicators such as covers, name plates and labels) in good conditions? ▸ How well are the waste, dust, oil and stain at acceptable level? ▸ Are looseness, vibration, wear, whinning noise and heat generation Good Points:	0	1 2 3 4	5 6	7 8	9 10
		Countermeasures (if ≤ 6) / OFI:				
2. Cleanliness of machine's peripherals & surrounding area	▸ How well is visual floor control (indicators such as covers, name plates and labels) in good conditions? ▸ How well are the waste, dust, oil and stain at acceptable level? ▸ Are looseness, vibration, wear, whinning noise and heat generation Good Points:	0	1 2 3 4	5 6	7 8	9 10
		Countermeasures (if ≤ 6) / OFI:				
3. Lubrication and tools/gauges conditions	▸ How well are the waste, dust, oil and stain at acceptable level? ▸ Are the tools, gauges and its indicators properly labelled and in good condition? Good Points:	0	1 2 3 4	5 6	7 8	9 10
		Countermeasures (if ≤ 6) / OFI:				
4. Immediate restoration of abnormality / undesirable conditions	▸ Is there any *fuguai* detected but not tagged by the team ? ▸ Is there any future plan to contain the forced deterioration exposed? Good Points:	0	1 2 3 4	5 6	7 8	9 10
		Countermeasures (if ≤ 6) / OFI:				
5. Identification of hard-to-access areas list; and generating sources	▸ Is there any list identifying the hard-to-access areas (HTAA) to ▸ Have improvements been made for covers and cleaning tools? Good Points:	0	1 2 3 4	5 6	7 8	9 10
		Countermeasures (if ≤ 6) / OFI:				
6. 5S workplace organization	▸ Are frequently used items within easy reach from operation area? ▸ Is there a duty roster for the area personnel for housekeeping ? Good Points:	0	1 2 3 4	5 6	7 8	9 10
		Countermeasures (if ≤ 6) / OFI:				
7. Understanding & application of TPM technique	▸ Do the related personnel understand TPM and participate in activities ? Is there any frequent formal meetings or communication ? ▸ Is there area related TPM education and training materials ? (i.e. area's OPLs, area's CLTM sheet, Tentative Standards) Good Points:	0	1 - 10	11 - 20	21 - 30	31 - 40
		Note: Number your score				
		Countermeasures (if ≤ 30) / OFI:				

Overall comments:
_____
_____

Final Judgement (tick '✓' where applicable)
☐ Unconditional Certification
☐ Conditional Cerification (upon submission of CAR)
☐ Re-Cerification (upon re-audit by TPM)

Fig. 4.9 Example checklist for AM audit

References

1. Ahmed S, Hassan M, Taha Z (2005) TPM can go beyond maintenance: excerpt from a case implementation. J Qual Maint Eng 11(1):19–42
2. Ahuja IPS, Kumar P (2009) A case study of total productive maintenance implementation at precision tube mills. J Qual Maint Eng 15(3):241–258
3. Bakri A, Rahim A, Rahman A, Noordin MY (2014) A review on the Total Productive Maintenance (TPM) conceptual framework. In: Applied Mechanics and Materials Vol. 660. Trans Tech Publications Ltd, pp 1043–1051
4. Gupta AK, Garg RK (2012) OEE Improvement by TPM Implementation: A Case Study. International Journal of IT, Engineering and Applied Sciences Research (IJIEASR) ISSN: http://www.2319-4413.org. Vol 1, No. 1
5. Ireland F, Dale BG (2001) A study of total productive maintenance implementation. Int J Prod Qual Manag 7(3):183–191
6. Nakajima S (1988) Introduction to TPM. Cambridge: Productivity Press
7. Steinbacher HR, Steinbacher NL (1993) TPM for America. Portland: Productivity Press
8. Suzuki T (1994) TPM in Process Industry. Portland: Productivity Press
9. Tsang AH, Chan PK (2000) TPM implementation in China: a case study. Int J Qual Reliab Manag 17(2):144–157
10. Wakjira W, Singh M (2012) Total productive maintenance: A case study in manufacturing industry. Global J Res Eng 12(1)
11. Yeomans M, Millington P (1997) Getting maintenance into TPM. Manuf Eng 76(4):170–173

Chapter 5
Risk-Based Inspection and Maintenance

5.1 Introduction

Modern business environment can tolerate no failures, especially involving unscheduled shutdowns, environmental contamination, individual injuries, and human fatalities. Thus, huge attempts must be in placed to control failure susceptibility at initial stage of design and operation. Once a system or plant is in service, maintenance is required more often than not to keep the operation safe and efficient. This is more necessary than ever in aging systems and structures, which may be operated beyond their intended design life. Major development in maintenance strategies can be seen in the last three decades in which the progress is encouraged by the expansion of volume, capacity, intricacy, and versatility of assets. Furthermore, the growing demands are vital in conformance of the environmental concerns and its relationship with maintenance management, the safety of personnel, the business viability, and the quality of products. In the 90, close relationship in between maintenance practices and failure incidences has been expressed by many researchers [1, 2]. Such failure is more often than not to cause detrimental consequences on the environment and may lead to major accidents.

The productivity of an organization is highly affected by downtime in terms of the production volume, cost of operation, and efficiency of customer services. This situation leads to unsatisfactory quality standard and further failures may affect safety and environment badly. According to Bevilacqua and Braglia [3], at least 15 and up to 70% of total production costs are represented by the quality of product, safety of plant, and the cost of maintenance. Major challenge of a maintenance engineer is to put up a maintenance strategy into operation which maximizes the availability and efficiency of equipment. The best maintenance strategy also controls the equipment deterioration rate and guarantees conformance of safety and environment in operation. Further, the total cost of operation can be minimized.

Maintenance concentration required for a system or an organization differs from one to another depending on the severity of operation and the criticality of the system. Some building may only need occasional painting and inspection of electrical and

© The Author(s), under exclusive license to Springer Nature Switzerland AG 2020
A. Bakri and Mohd. A.-F. Mohd Szali Januddi, *Systematic Industrial Maintenance to Boost the Quality Management Programs*, SpringerBriefs in Applied Sciences and Technology,
https://doi.org/10.1007/978-3-030-46586-5_5

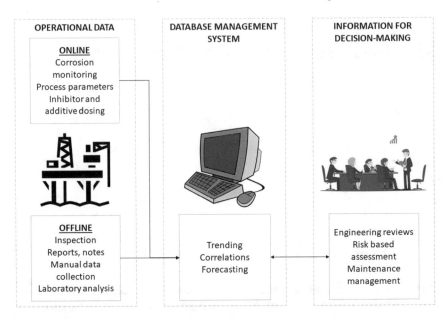

Fig. 5.1 Combined data from various sources of a facility to be integrated into management information for maintenance system

plumbing lines, but some other may need extensive maintenance scheduling such as in chemical plants, aircraft, marine equipment, power generation plant, etc. Figure 5.1 shows the integrated inspection and monitoring programs in a facility to produce management data. This data can be used for decision-making in maintenance system of the facility.

Maintenance system has been conventionally practicing reactive or blindly proactive strategies. A component is only replaced or fixed once it can no longer be functioning. This maintenance strategy known as the corrective maintenance is undesirable due to extensive labor service, retain catastrophic failures and promote unnecessary maintenance. On top of that, the cost is further elevated by maintenance labor and downtime, as well as cost associated with safety concern and customer satisfaction. Preventive maintenance on the other hand is lack of information such as the "fitness" of a particular component. The fitness if wrongly assessed especially for safety-critical components may lead to disasters. This method involves fixing or replacing failed equipment to prevent reoccurrences of such failure. Over the years, the maintenance management techniques have been progressively developed from here to the implementation of condition monitoring, reliability-centered maintenance, and expert systems (Fig. 5.2). Currently, risk-based maintenance methodologies are beginning to come forward and gather more interests.

The development of maintenance philosophies can also be shown as in Fig. 5.3, in which the policy evolution is categorized as the first, second, third, and recent generation. The first generation is referred to the period before World War II, when

Fig. 5.2 The evolution of maintenance strategy

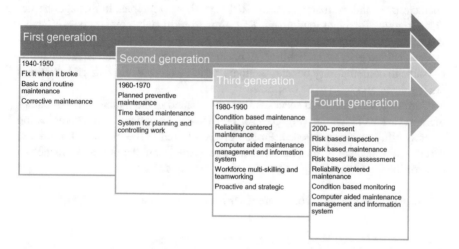

Fig. 5.3 The development of maintenance philosophies

most industries were poorly mechanized. Simple equipment is redesigned from time to time to improve reliability and repair procedure. Machines were usually kept in used until failure occurs with no regards for the possibility of its failure.

Before 1980s, greater dependency on machineries was experienced. Industries have been more complex than ever tipping the maintenance cost over the other relative operational costs. During this generation, the maintenance practice was commonly known to encompass excessive treatments interfering with normal operations and stimulating crashes as the operations were overlooked. Toward year 2000 thereafter, the third generation maintenance policies demand for automation, efficient production system, as well as standardized product and service quality. Condition-based maintenance, reliability-centered maintenance, and computer-aided maintenance management system were started to be incorporated in practice in order to satisfy those demands.

Beyond the year 2000, inception of risk-based inspection and maintenance started to be implemented on top of RCM and CBM. As safety and maintenance are not mutually exclusive in terms of function and implementation, integration of both in optimizing plant capacity seems to promote profitability and improve the total life-cycle cost. In this regards, safety and environmental concerns are not found in the middle ground.

The industry was moving toward better risk control at reasonable cost since the 1990s. A strategy for maintenance scheduling was proposed by Chen and Toyoda, in which the increase risk of failure in an equipment or machine is made even and stan-dardized [4]. Optimum maintenance schedule can be developed by combined effects of reliability approach and risk assessment strategy or standardization. This method, known as risk-based maintenance (RBI) [5], is scheduled based on the formulated likelihood of failure scenarios. RBI implemented risk analysis, which identifies, characterizes, and evaluates hazards. RBI uses these principles to manage inspec-tion programs for plant equipment. RBI aims to ultimately create a cost-effective inspection and maintenance system which at the same time guaranteed appropriate mechanical integrity and reliability of concerned facilities. This unmistakably indis-pensable in contemporary business settings, where technical and financial resources are regulated and have to be optimized.

Currently, RBI has been successfully implemented in the maintenance of equip-ment in various industries (Fig. 5.4) [6, 7]. Risk can be defined as the combination of probability and consequence. Probability is referred to the likelihood of failure, and the consequence is the quantified damage ensued from the failure. Damage can be described in terms of injury, fatality, and property damage.

$$RISK = \text{likelihood of failure (LOF)} \times \text{consequence of failure (COF)}$$

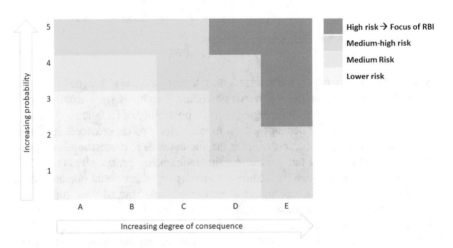

Fig. 5.4 Risk levels computed from the likelihood of failure and its consequences. RBI focus is given on the items with higher risk

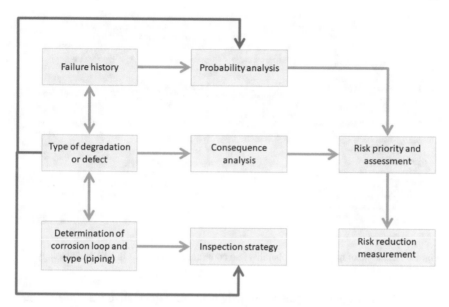

Fig. 5.5 General RBI method used in industries. For piping system, the workflow usually involves determination of corrosion loop as per API-581

Thus, the probability analysis and consequence analysis in Fig. 5.5 are the most important assessment in RBI methodology, which may be presented by hazard, environmental factors, and production loss. To do this, there are three analytical levels of RBI analysis; (1) qualitative, (2) semi-quantitative, and (3) detailed quantitative.

The calculated probability of failure is compared with known criteria to obtain the frequency of maintenance should be delivered to be able to minimize the risk of failure. RBI aims to cope with meaningful risk while exploiting limited resources. It is well known that in many practical settings, at least 80% of risk exposures are dominated by only 20% of equipment items [8]. Illustrated in Fig. x is the outcome of increased probability and high degree of consequence that contributes to elevated risk level. Risk levels based on the equipment operational system are established and ranked accordingly in RBI for systematic analysis. Thus, when the inspection is found to be reasonably low, cutting down the risk levels and further trimmed down the cost of maintenance work associated with it.

A simple workbook can be used to audit LOF and COF qualitatively. The assessment can be evaluated from recognized provoking probability factors such as the quantity of equipment available, the damage mechanisms (corrosion mechanism, fatigue cracking, etc.), the inspection efficiencies, current condition of the equipment, as well as the operational system and design of equipment. On the other hand, the consequence factors consider potential hazards and its total cost. This concept is briefly demonstrated in Fig. 5.6. Risk assessment resulted from these analyses can be used to locate concerning area of interest and to determine which of these areas call for maximum inspection consideration and what not.

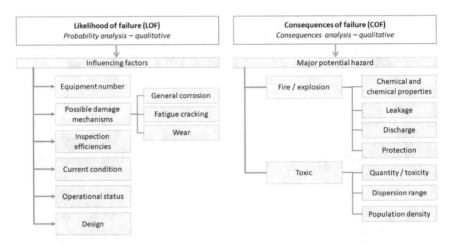

Fig. 5.6 Example of qualitative probability and consequences analysis in RBI methodology

In order to avoid discrepancies in judgment from qualitative assessment, semi-quantitative method should be taken into account for better risk assessment results. In quantitative method, the risk levels are established from potential losses. The LOF of an equipment is calculated based on the generic failure frequency (GFF). GFF is defined as the failure histories of a particular equipment in industries multiplied by the equipment modification factor and management evaluation factor. COF assessment in quantitative method is based on the calculated losses, for example, the hazard, environment, downtime duration, maintenance expenses, etc. The quantitative analysis of LOF and COF is briefly shown in Fig. 5.7. Based on this calculation, rick exposure of a particular equipment can be directly assessed.

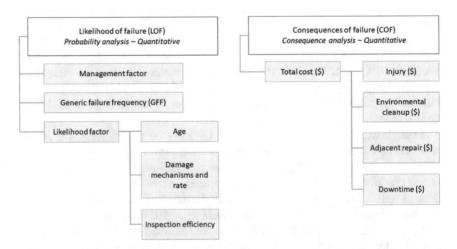

Fig. 5.7 Example of quantitative probability and consequences analysis in RBI methodology

API RBI program outlines three approaches to reduce risk;

1. Inspection and monitoring strategy optimization
2. Materials of construction alternatives
3. Key process parameters control

The shift of over-inspection low-risk items and of under-inspection high-risk items can be done by reviewing and modifying the inspection and monitoring plan. On top of that, all relevant damage mechanisms recognized in the RBI program must be addressed by appropriately selected inspection techniques. Data of the failure rate for different materials and items can be implemented to rummage around for alternatives materials or items. This would rather be done rationally than simply giving in onto the intent of cost minimization. Further, the key process parameters which influencing the damage can be evidently acknowledged. Subsequent monitoring on these parameters makes certain of the damage to be within the safe operation windows. Changes in any process and its effects on the failure risk can be assessed wisely.

Poor maintenance practices may result in reduced capacity of production as downtime increases and uptime underperforming assets. Further, the underperformed assets incur real and opportunity cost penalties. Products and services quality will then be fell off, affecting customers' satisfaction and may contribute to the loss of sales. In terms of safety hazard, failures may jeopardize the life of personnel and surrounding community, as well as suffering further financial losses.

5.2 Proposed RBM Methodology by Khan et al. [9, 10]

Maintenance tasks in an organization can be planned and strategized by using risk factor as a general methodology that can be implemented onto all types of assets. In this case, the risk factor needs be quantified accurately, in which its accuracy hinges on the quality of consequences study and estimation of failure probability. Risk can be defined as follows;

$$\text{Risk} = \text{failure probability} \times \text{failure consequences}$$

Using data of failure and the related economic consequences, risk-based approach maintenance strategy can be developed to minimize hazards due to unexpected failure of equipment. For this objective to be met, risk analysis to identify, characterize, quantify, and evaluate the loss from failure event should be addressed. From the risk analysis, the factors and the origin of the failure can be identified. Further, the mechanism of the identified factors to initiate and advance the failure. The analysis also should be able to quantify the likelihood of failure reoccurrence and acknowledge its consequences.

Quantitative risk assessment usually resulted in cost impact per unit time. The computed cost helps in prioritizing the items which have been assessed. Using fault trees, a particular event cycle can be mapped on specific consequence. On

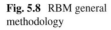

Fig. 5.8 RBM general methodology

the other hand, qualitative assessment is more conservative represented by a risk matrix. However, the value presented in the risk matrix is only a relative number which cannot be accountable beyond the framework of the matrix. Thus, prioritization based on the value is always open to question as it is very subjective in nature. In overall, RBM aims to reduce the risk by and large of an operating facility. High- and medium-risk areas will need focused maintenance effort, while low-risk area entails for minimum maintenance effort. This will optimize total scope of maintenance work and cost.

In risk-based maintenance (RBM), reduction of failure frequency and consequences are expected in which the maintenance planning and decision-making are improved. With minimal cost and safety constraints, RBM adapts new strategies integrating dynamic evolution model and risk assessment. RBM integrates reliability approach and risk assessment strategy to achieve optimum maintenance schedule. Maintenance effort is focused more on high and medium risk, whereas maintenance effort is minimized in low-risk areas. The number of preventive tasks required is suggested through RBM. To initiate the RBM process, the complete system under study is first divided into small manageable units (Fig. 5.8), which is separated into three modules; (1) risk determination, (2) risk evaluation, and (3) maintenance planning.

5.2.1   Risk Estimation

This first module consists of identification and estimation of risk in four steps as shown in Fig. 5.9. In the first step, failure scenario will be developed from the series of events which are recognized to lead and cause failure in a particular system. Most of the time, failure is observed to be initiated by a combination of sequential events. Failure scenario developed here will be the foundation of the risk study which gives information of the expected circumstances. This way, measures to deter occurrences of failure can be established. In 2001, the author established maximum

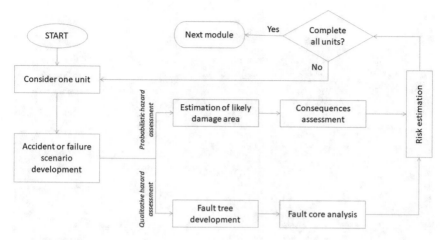

Fig. 5.9 Risk estimation module

credible accident scenario (MCAS) as a systematic procedure to evaluate failure in a particular system [11]. From MCAS, a short list of failure scenarios most relevant to the system is formed.

Based on the failure scenario, the equipment will then be prioritized accordingly. Considering the failure event takes place, consequence analysis is needed to be done. First is to quantify the consequences based on the possible damage area radius, the possible damage on the property, and the toxic effects. Information from this damage radius will then be used to evaluate the damage impact on human health, environment, and production. Figure 5.10 shows the workflow of this procedure which involves various mathematical models [12].

Frequency of failure is determined from failure data and human reliability data by using fault tree analysis. In the year 2000, the author had also suggested a method known as analytical simulation to develop the fault tree probability study [13]. The

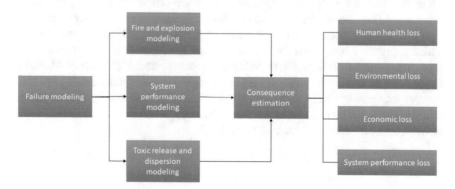

Fig. 5.10 Consequence assessment chart

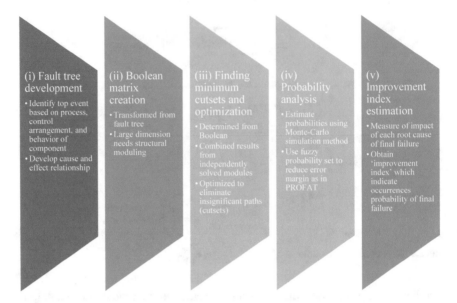

Fig. 5.11 Fault tree probability study procedure, in which large dimensional Boolean matrix is restructured [14] and the probability error margin can be reduced using PROFAT [15]

study consists of five important steps; (1) the fault tree development, (2) Boolean matrix creation, (3) the finding of minimum cut-sets and optimization, (4) the probability analysis, and (5) the improvement of index estimation (Fig. 5.11). The risk will be finalized from the results of the consequence and failure probability.

5.2.2   Risk Evaluation

Figure 5.7 shows the algorithm used to evaluate the estimated risk once the risk is determined from the aforementioned procedure. The first step of this algorithm is to set up acceptance criteria by identification of specific risk. The author used open-ended approach which allow for risk acceptance level based on system nature and type. Acceptance criteria available included as low as reasonably possible (ALARP), Dutch acceptance criteria, and USEPA [16] is applied to the estimated risk. Improved maintenance plan will be needed by unit with estimated risk exceeding the acceptance criteria. These units will be studied comprehensively in order to reduce the risk level through better maintenance plan (see Fig. 5.12).

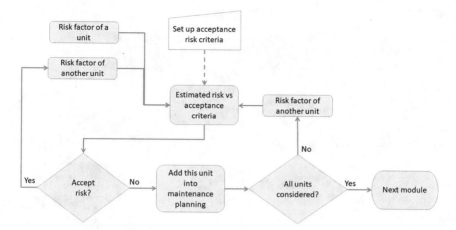

Fig. 5.12 Risk evaluation module

5.2.3 Maintenance Planning

Each cause of failure is reevaluated to determine which impacted the failure probability more than the others. This is done by reverse fault tree analysis, in which the probability failure value is verified and the maintenance plan is established. The established maintenance plan will then be verified to produce acceptable total risk level.

The objective of maintenance is to increase the availability of a system which reliable in terms of safety and environmental concerns. At the same time, maintenance is also expected to optimize the total life-cycle cost. RBM can address five important questions of a system running durably and faulty-free, as follows;

1. What are the factors contributing to system failure?
2. How are the factors causing failure in the system?
3. What is the probable consequences when the failure occurs?
4. What is the probability of the failure?
5. How frequent the inspection/maintenance of the concerning equipment which contribute to such failure?

Answering these questions through RBM implementation will bring about better asset and capital utilization. Furthermore, the implementation of RBM can improve existing maintenance policy by optimizing judgment and evaluation process throughout the life cycle of a concerning system (Fig. 5.13).

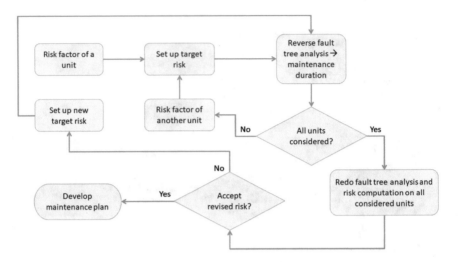

Fig. 5.13 Maintenance planning module

5.3 RBM for Medical Devices

Emerging technologies in healthcare industry provides promising diagnostic and therapeutic possibilities but such enhancement in frequency and complexity of system interventions may increases the risk of failures. Failures in medical devices may be harmful to patients and poses risk of injury to clinical staff from simple, direct hazards such as accidental contact with poorly insulated parts of instruments, or mechanical failures in scanners and pumps. Thus, effective maintenance system which prioritizes the medical devices based on key criteria is needed to be implemented. To date, many hospital engineering departments have started to consider risk as requirement for maintenance activities of medical device. However, current practice misses some important criteria [17] such as the number of patients served by a particular device, the economic loss, the mean time to repair, the use-related hazards, etc. Furthermore, some devices are assigned with the same level of risk despite diversified operational and environmental conditions of the hospital. Such problems lead to misclassification of device and affected the maintenance activities should be implemented onto it. Some important aspects in RBM of medical devices in healthcare industry are left overlooked and demanded for improvement [18].

Assessment of medical devices is considered to be multi-criteria decision-making problem, which require intervention of different expert's evaluation. Prioritization of medical devices according to conventional criteria must be reassessed, with some new criteria to be considered. Criteria and prioritization tables used should be presented with modesty to be understood by all, avoiding misconception. Further, consideration on uncertainties must be given as pointed out by the experts, and a comprehensive and systematic framework to prioritize maintenance activities for medical devices is traditionally missed out. Thus in 2015, Afshin Jamshidi et al. presented a novel

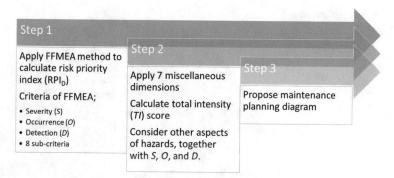

Fig. 5.14 Proposed three-step approach for clinical engineers to prioritize medical devices for best maintenance strategy [18]

fuzzy multi-criteria decision-making approach to prioritize medical device within an established RBM framework [18]. The proposed three-step approach had been proven to do revisiting and reassessment of medical device risk score criteria (Fig. 5.14).

The proposed approach is based on the three key features of RBM strategy: personnel safety effect, environmental threat, and economic loss (Fig. x). The economic loss can be directly measured in terms of money, but the other two features are fairly subjective. Personnel injury and pollution are too complex to be assigned with dollar sign. However, the impact of these two features can be expressed in terms of process performance. Thus, the authors chose to implement failure modes and effect analysis (FMEA) to identify and assess the impact by forming a semi-quantitative method based on the subjective information obtained from pass experience.

To calculate the probability of failure of the targeted facility, the author used Weibull distribution model. The risk was then computed by multiplication of the failure with the consequences. The concept of risk index was evaluated by weighing the calculated risk against the known acceptable risk criterion, in which resulted in the three risk indexes. The relative importance of the three consequences features (as shown in Fig. 5.15) was determined using an analytic hierarchy process, resulted in a weight factor which was used to integrate the three risk indexes. Under this computed risk constraint, proper maintenance schedule was finalized.

Fig. 5.15 Key features in
RBM strategy

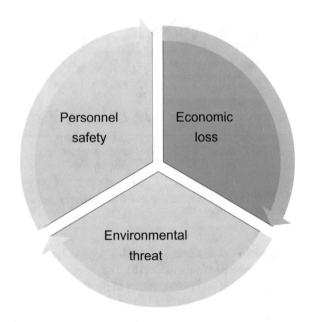

References

1. Kletz T, Amyotte P (2019) What went wrong?: case histories of process plant disasters and how they could have been avoided. Butterworth-Heinemann
2. Kumar U (1998) Maintenance strategies for mechanized and automated mining systems: a reliability and risk analysis based approach. J Mines Met Fuels 46:343–347
3. Bevilacqua M, Braglia M (2000) The analytic hierarchy process applied to maintenance strategy selection. Reliab Eng Syst Saf 70:71–83
4. Chen LN, Toyoda J (1989) Maintenance scheduling based on two level hierarchical structure to equalize incremental risk. In: Conference papers power industry computer application conference (IEEE), pp 431–437
5. Inspection R-B (2000) Base resource document. API Publications 581
6. Chang M, Chang R, Shu C, Lin K (2005) Application of risk based inspection in refinery and processing piping. **18** 397–402
7. Ridgway M (2001) Classifying medical devices according to their maintenance sensitivity: a practical, risk-based approach to PM program management. Biomed Instrum Technol 35:167–176
8. Lee C, Teo Y (2001) Det Norske Veritas RBI study for the ammonia storage plant. Fertil. Kaohsiung Ammon. Termin, ROC Taiwan
9. Khan FI, Haddara MM (2003) Risk-based maintenance (RBM): a quantitative approach for maintenance/inspection scheduling and planning. J Loss Prev Process Ind 16:561–573
10. Khan FI, Haddara M (2004) Risk-Based Maintenance (RBM): a new approach for process plant inspection and maintenance. Process Saf Prog 23:252–265
11. Khan FI (2001) Use maximum-credible accident scenarios for realistic and reliable risk assessment. Chem Eng Prog 97:56–64
12. Khan FI, Abbasi SA (1998) MAXCRED—A new software package for rapid risk assessment in chemical process industries. Environ Model Softw 14:11–25
13. Khan FI, Abbasi SA (2000) Analytical simulation and PROFAT II: a new methodology and a computer automated tool for fault tree analysis in chemical process industries. J Hazard Mater 75:1–27

14. Shafaghi A (1988) Structure modeling of process systems for risk and reliability analysis. Kandel, Avni, pp 45–64
15. Fadt P, Analysis T (1999) PROFAT: a user friendly system for probabilistic fault tree analysis. Process Saf Prog 18:42–49
16. Lees F (2012) Lees' loss prevention in the process industries: hazard identification, assessment and control. Elsevier
17. Rice WP (2007) Medical device risk based evaluation and maintenance using fault tree analysis
18. Jamshidi A, Abbasgholizadeh S, Ait-kadi D, Ruiz A (2015) A comprehensive fuzzy risk-based maintenance framework for prioritization of medical devices. Appl Soft Comput J 32:322–334

Chapter 6
Computerized-Based Maintenance

Over 50 years ago, the very first computerized maintenance management systems (CMMS) software was first developed. This started with the math-based system design with objective to improve documentation, standardizing, and verification of manufacturing processes. In 1965, some of the largest manufacturing firms had implemented CMMS software in their maintenance practice. The CMMS software was later evolved and expanded ever since, in parallel with enhancement in computing power and the development of Internet. This technology advancement had made the CMMS software to be accessible and convenient to small- and medium-sized enterprises all over the world.

Throughout the history, the CMMS evolution can be subdivided into four generations (Fig. 6.1). The first generation utilized the concept of punch cards, in which the maintenance technicians were reminded to perform recurring maintenance tasks. The work-order data would be punched in onto the cards to be read by the computer. The system was based on programming languages such as Fortran and Cobol, which ran on centralized IBM mainframe computers. This kind of system was only affordable and accessible to large asset-intensive businesses.

Ten years later, the punch card was replaced by paper forms; thus, the work orders were printed out on paper before circulated to maintenance personnel by hand. Completed work was recorded in the same form and returned to the data entry clerk to be keyed in into the mainframe computer. At this point of time, implementation and the use of CMMS is a huge investment for an organization, thus became an unaffordable luxury to many small- and medium-sized businesses. However, with the invention of mini computers in the 1980s, the CMMS software implementation could then be stretched to all. After a particular work order was completed, the maintenance technicians would enter data on the plant green-screen terminals. These terminals evolved later from year to year becoming more sophisticated to incorporate more functions, such as reporting. After the 1990s, homegrown Microsoft Access-based CMMS applications were a hit in many organizations. This is due to the availability of personal computers (PCs) and the advancement of networking capacity.

© The Author(s), under exclusive license to Springer Nature Switzerland AG 2020
A. Bakri and Mohd. A.-F. Mohd Szali Januddi, *Systematic Industrial Maintenance to Boost the Quality Management Programs*, SpringerBriefs in Applied Sciences and Technology,
https://doi.org/10.1007/978-3-030-46586-5_6

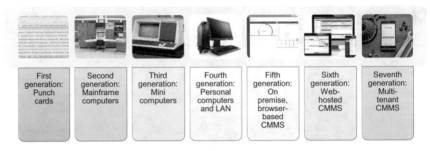

| First generation: Punch cards | Second generation: Mainframe computers | Third generation: Mini computers | Fourth generation: Personal computers and LAN | Fifth generation: On premise, browser-based CMMS | Sixth generation: Web-hosted CMMS | Seventh generation: Multi-tenant CMMS |

Fig. 6.1 The evolution of CMMS throughout the history

The applications were then served as the basis of new, advanced CMMS software businesses.

Microsoft Access-based CMMS requires to be set up on each terminal for retrieval. However, there are numbers of downsides of this system which includes limited history of work order, disintegrated database, and no trails of expenditure. Furthermore, the system itself is scattered thus requires printed documentation for access, and upgrading the system incurring huge cost. Therefore, just before Y2K, most vendors developing CMMS software migrated the system from Microsoft Access-based to browser-based applications. The CMMS developed at this point was more accessible, but appeared to be outdated—more like a terminal interface for mainframe computers. The update process can be done remotely by the vendors but must be monitored locally as the software ran on local servers, which bring down the CMMS for extended period.

As the Internet started to emerge and giving more access beyond the company's firewall, many vendors then started to develop web-hosted CMMS applications. Access via Internet can be done through any Internet-connected devices such as laptops, tablets, and smartphones. Scheduling can be done efficiently, work orders can be monitored remotely, and reporting can be more conservative. Vendors have full ownership as the CMMS runs on vendor's server, and any updates and upgrades are done by vendors. However, in case of server failure, more time is required to get each system to relaunch on new server from the most recent backup. Nowadays, most CMMS software development are embracing the cloud computing service in order to eliminate the need for complex server installation and configuration. The environment allows customization on modules without expensive consultation service. Furthermore, the cloud system itself is reliable, secure, and affordable. Failure on one hardware will not affect the other devices in synced as the application is stored on virtual server. At predetermined intervals, upgrades and updates happen automatically. This multi-tenant technology conveys the CMMS software to be low in cost thus affordable to all regardless of the business size.

6.1 Commercial CMMS or Self-established CMMS?

A particular company usually has already computerized the maintenance work using basic software. However, in the event of acquiring prim and proper CMMS, a company should be considering the size of the system and the network to be expended. Many maintenance groups choose to have personal computer network which is separated from the company wide system thus allows for manipulation of information and bigger database. This would require in-depth expertise of the CMMS vendor and high dependency on the efficiency of the company's IT support. However, the important question on this matter remains; whether the company wants to pursue on the commercially available CMMS or self-develop the system as a whole?

Before we answer this question, let us take a look at the current practice of the maintenance group of this company without a formal CMMS. At the moment, with the aid of internet, communication between personnel is made efficient with emails. Maintenance work can be done easily by contacting manufacturers or vendors which may put up the maintenance procedure online. These manufacturers or vendors can also be reached out by emails and phone calls. Further, searching for datasheet is feasible and having the on-site records can be done faster. Purchase of materials can also be effectively done online. The company may also have a set of database recording all equipment's details and information. Simple spreadsheet and word processor may be used to print and record all preventive maintenance work order. Improvement on this system can be made economically using word processor and filing system.

Therefore, no company should ever pursue on acquiring CMMS for the sole reason of automation and single clerk replacement. Nevertheless, here we also highlight the benefit of using prim and proper CMMS which can be taken into consideration.

6.2 CMMS Selection

The main factor of consideration when picking up the CMMS is the final cost it incurred. The right CMMS must not only perform the tasks appropriately but also would not be stressing on the budget. Usually, the first step is to recognize and classify the required features and the optional or nice-to-have features (see Fig. 6.2). For example, a hospital may need an accurate record of inventory, maintenance history, equipment operating hours, and critical reporting [1]. Further consideration should be given in the selection of vendor. Vendors are preferred to be in the business for quite some times with specialization in CMMS. It is suggested that a particular organization to spread the words among the other companies in the same industries of having the market open for a CMMS through meetings, memo, or word of mouth. This will attract vendors and increase the competition among them, which on the other hand would bring the price down and bring up the number of features to be offered in the package.

Fig. 6.2 Example of required and optional features of CMMS in a hospital setting [1]

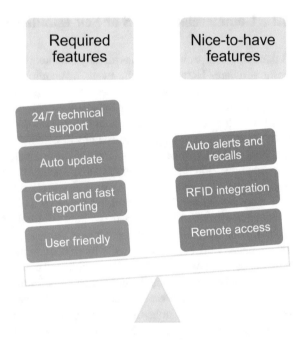

Different systems have different strengths and weaknesses. Thus, to find one system fits all is very challenging. The CMMS may be very user-friendly, but lack of technical support and dynamics, such as CMMS built on MS Access Database. The main weakness of such system is that it does not incorporate much of advanced features such as construction of temporary tables or have the user-defined functions, views, and procedures. In comparison to MS SQL or MySQL, the database is much smaller with limited concurrent users of up to only 255 users and support constrained quantity of objects. While the end-user may not have problems with this capacity, the IT personnel whom would be working with the database may find these restrictions frustrating.

Interviews and analysis of the current system in an organization must be made to diagnose the needs and the necessary features for a new system. The point of improvement would supposedly take in consideration of the best practices suggested in the literature in combination of the organization needs. The needs of an organization may include, among others, the maintenance planning, scheduling, performance measure and improvement, and spare parts management. With the support from the information system of the organization, the maintenance strategy can be improved from the visualization of maintenance tasks. This involves the overview of each equipment or machine status as well as the Gantt chart of maintenance planning and scheduling associated to each equipment and machine. Further, real-time monitoring and alerts for machines and equipment can be set in accordance to the parameter deviation from a specific value range for predictive maintenance. Then, the intervention periods, parameters, and limits of predictive maintenance will be recorded in the

CMMS, tailoring the information from suppliers and the experience of maintenance technicians.

An efficient CMMS not only include planning of maintenance work but also incorporate scheduling in the form of Gantt diagram. The maintenance technician will be able to confirm each and every assigned maintenance intervention on him on weekly or monthly basis. However, this scheduling system is very much limited to consider subjective standpoint of the maintenance technician such as the technician's experience, availability, skills, inventory, and other overlapping issues. In conclusion, the selection of CMMS must include consideration and expertise from IT department. This is crucial in order to understand the capacity of the company server, database requirement, network infrastructure, as well as designing user-defined reports. Selecting and establishing a CMMS system in the company would give new experience and knowledge to the person in charge especially the maintenance team. Thus, it is important that the information to be shared among other companies within the same industry for further development and enhancement of maintenance management.

6.3 Benefits of CMMS Implementation

In 1998, Christer Idhammar reported that the implementation of CMMS has success rate of only 18%. This may conclude that CMMS is not the ultimate solution for maintenance issues in a plant or a company. Rather, CMMS is actually a tool for the whole organization—not only the maintenance staff—to provide information and database of maintenance related issues. Thus, in many organizations, CMMS system is best known as the assets management system. This is to make the system to benefit beyond the maintenance department and be usable to everyone in the organization. CMMS provides information in order to facilitate the practice of maintenance management. Maintenance management in an organization aims in minimizing the labor cost of maintenance and the cost of maintenance-related materials. As maintenance management works in parallel with the operation of the organization, maintenance management is at the same time curtailing the losses in production.

However, with the abundance of CMMS available in the market, the selection of the most appropriate system to a particular organization is a real challenge for many. It is commonly recognized that the presented system available do not satisfy its function and purpose due to the lack of the designer's experience in maintenance management [*Handbook of Maintenance Management, F Herbaty*]. Such system will be needed modification from once in a while to accommodate the users' demand. Further, the system will have the need for periodic evaluation in order to make certain of its functionality and reliability. Thus, many organizations are oftentimes baffled to either self-developing the CMMS or purchase from those commercially available. Implementation of CMMS in an organization is beneficial to the maintenance practice by automation and facilitation of current practice which will eventually improve the

Fig. 6.3 Basic functions of
CMMS

maintenance efficiency. In general, computerization is mostly reliable and cost effective (Fig. 6.3). This starts with standardization of maintenance work. Standardization improves consistency, and consistency improves reliability. However, standardizing workflow can be both beneficial and disastrous.

With the advancement in information and communication technologies, the function of CMMS in an organization can be made better in terms of efficiency and effectiveness. Many experts agree that a CMMS supports maintenance strategy by a set of functions and applications which include [2–4];

- Management of assets consists of record of all assets (or equipment and spare parts) with details on maintenance history and part list;
- Management of work order to set and release the work order to the assigned maintenance technicians;
- Management of preventive maintenance to support maintenance planning, scheduling, and control of activities;
- Management of reports, in which the CMMS is able to process large amount of data and analyze the data to indicate performance over time.

However, reviews on the commercially available CMMS signify some weaknesses and limitations [5, 6]. The system can be improved by including condition monitoring data analysis function, incorporate the failure diagnostic tools, support on resource allocation, and decision analysis function. Furthermore, many commercially available CMMS are unspecified, non-tally to the organization in particular. Thus, many organizations choose to self-developed the CMMS instead.

New emerging concept in maintenance technology recently is the e-maintenance (Fig. 6.4), in which information and communication technology is formalized and disciplined through the entire life cycle [7]. There are two main factors attributed to the emergence of e-maintenance concept in maintenance system. The first factor is the authority from e-technologies which allows for speedy, efficient, and proactive maintenance. This authority enhances the maintenance workflow in an organization. The second factor is the demand of business performance integration, which allows for open collaboration with other services of the e-enterprise. In short, e-maintenance can benefit maintenance strategy in terms of maintenance types and strategies, support and tools, as well as maintenance activities (Fig. 6.5).

As a maintenance strategy, e-maintenance controls maintenance tasks electronically using instantaneous data obtained from advanced equipment and machines [9]. The purposes of e-maintenance include maintenance documentation record, immediate access to information, remote data gathering to determine key performance index, and integration of the maintenance system with other information systems [10]. The integration and synchronization of maintenance system with other reliability systems make instantaneous information on assets accessible at any time [11, 12].

E-maintenance can also be a part of maintenance plan. The future e-automation manufacturing demands to consider computerized and systematized approaches of CBM, proactive and collaborative maintenance, remote maintenance and service support, provision for instantaneous information, as well as integration of production with maintenance [13]. In order to put the e-maintenance plan into practice, a

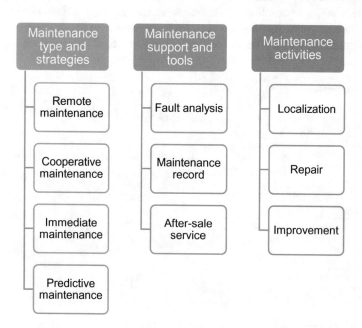

Fig. 6.4 Advantages of e-maintenance [8]

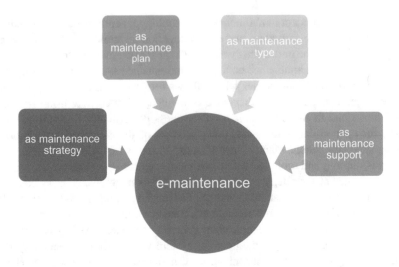

Fig. 6.5 E-maintenance concept which integrates information and communication technologies

proactive e-maintenance organization is required to be versatile which incorporate monitoring, diagnosis, prognosis, as well as decision and control process. In general, e-maintenance is a gradual substitution representation of the traditional maintenance (Fig. 6.6).

Zhang et al. [16] in their study on service platform in e-maintenance regard the e-maintenance as a consolidation of Web service technology and agent technology. This unification offers intelligent and cooperative features to be established in industrial automation system. This concept of e-maintenance as maintenance support is

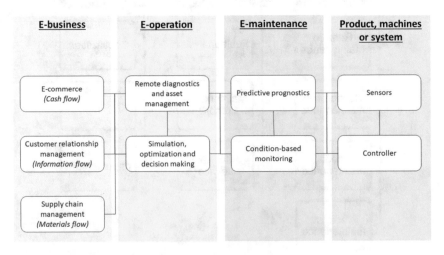

Fig. 6.6 E-maintenance from enterprise's perspective [14, 15]

also supported by the work of Marquez and Gupta later in 2006 [17], in which the e-maintenance is defined as a distributed intelligent environment. The environment was described to be consisted of information processing capability, decision support and communication tools, as well as the collaboration in between maintenance processes and expert systems. In 2008, Muller et al. [8] proposed an e-maintenance definition which had consider the maintenance terminology from European Standard (EN 13306:2001) and the concept of e-maintenance as an e-manufacturing component [18, 19]. Based on this, e-maintenance is defined as;

Maintenance support which includes the resources, services and management necessary to enable proactive decision process execution. This support includes e-technologies (i.e. ICT, Web-based, tether-free, wireless, infotronics technologies) but also, e-maintenance activities (operations or processes) such as e-monitoring, e-diagnosis, e-prognosis, etc.

References

1. Ilir K (2008) Selecting a computerized maintenance management system Clin Eng Manag 276–8
2. Cato WW, Mobley RK (2001) Computer-managed maintenance systems: a step-by-step guide to effective management of maintenance, labor, and inventory. Elsevier
3. Donoghue CDO, Prendergast JG (2004) Implementation and benefits of introducing a computerised maintenance management system into a textile manufacturing company 154: 226–32
4. Zhang Z, Li Z, Huo Z (2006) CMMS and its application in power system. Int J Power Energy Syst 26:75
5. Rastegari A, Mobin M (2016) Maintenance decision making, supported by computerized maintenance management system. In: 2016 annual reliability and maintainability symposium (RAMS), IEEE, pp 1–8
6. Labib AW (2004) A decision analysis model for maintenance policy selection using a CMMS. J Qual Maint Eng
7. Candell O, Karim R, So P (2009) Robotics and Computer-Integrated Manufacturing eMaintenance—Information logistics for maintenance support 25: 937–44
8. Muller A, Crespo A (2008) On the concept of e-maintenance : review and current research 93: 1165–87
9. Tsang AHC (2002) Strategic dimensions of maintenance management. J Qual Maint Eng 8:7–39
10. Wireman T (2004) Total productive maintenance. Industrial Press
11. Baldwin RC (2001) Enabling an e-Maintenance infrastructure. Maint Technol 12:408–429
12. Moore WJ, Starr AG (2006) An intelligent maintenance system for continuous cost-based prioritisation of maintenance activities. Comput Ind 57:595–606
13. Ucar M, Qiu RG (2005) E-maintenance in support of e-automated manufacturing systems. J Chinese Inst Ind Eng 22:1–10
14. Koc M, Lee J (2001) A system framework for next-generation E-maintenance systems. China Mech Eng 5:14
15. Lee J, Ni J (2004) Infotronics-based intelligent maintenance system and its impacts to closed-loop product life cycle systems. In: Proceedings of international conference on intelligent maintenance systems, Arles, France, pp 15–7
16. Zhang W, Halang W, Diedrich C (2003) An agent-based platform for service integration in E-maintenance 426–33

17. Crespo A, Gupta JND (2006) Contemporary maintenance management : process, framework and supporting pillars 34: 313–26
18. Lee J (2003) E-manufacturing—fundamental, tools, and transformation 19: 501–7
19. Muller A (2004) Copyright © IFAC information control problems in manufacturing, Salvador, Brazli, pp 359–65

Printed in the United States
By Bookmasters